An Introduction to the Event-Related Potential Technique

Cognitive Neuroscience
Michael S. Gazzaniga, editor

An Introduction to the Event-Related Potential Technique

Steven J. Luck

The MIT Press
Cambridge, Massachusetts
London, England

MIT Press books may be purchased at special quantity discounts for business or sales promotional use. For information, please email special_sales@mitpress.mit.edu or write to Special Sales Department, The MIT Press, 55 Hayward Street, Cambridge, MA 02142.

This book was set in Melior and Helvetica on 3B2 by Asco Typesetters, Hong Kong. Printed and bound in the United States of America.

Library of Congress Cataloging-in-Publication Data

Luck, Steven J.
An introduction to the event-related potential technique / Steven J. Luck.
    p.   cm. — (Cognitive neuroscience)
Includes bibliographical references and index.
ISBN 0-262-12277-4 (alk. paper) — ISBN 0-262-62196-7 (pbk. : alk. paper)
1. Evoked potentials (Electrophysiology) I. Title. II. Series.
QP376.5.L83   2005
616.8′047547—dc22   2005042810

10 9 8 7 6 5 4 3 2

Contents

Preface

The event-related potential (ERP) technique has been around for decades, but it still seems to be growing in popularity. In the 1960s and 1970s, most ERP researchers were trained in neuroscience-oriented laboratories with a long history of human and animal electrophysiological research. With the rise of cognitive neuroscience and the decline of computer prices in the 1980s, however, many people with no previous experience in electrophysiology began setting up their own ERP labs. This was an important trend, because these researchers brought considerable expertise from other areas of science and began applying ERPs to a broader range of issues. However, they did not benefit from the decades of experience that had accumulated in the long-standing electrophysiology laboratories. In addition, many standard ERP techniques are often taken for granted because they were worked out in the 1960s and 1970s, so new ERP researchers often do not learn the reasons why a given method is used (e.g., why we use tin or silver/silver-chloride electrodes).

I was fortunate to be trained in Steve Hillyard's lab at University of California, San Diego, which has a tradition of human electrophysiological research that goes back to some of the first human ERP recordings. My goal in writing this book was to summarize the accumulated body of ERP theory and practice that permeated the Hillyard lab, along with a few ideas of my own, so that this information would be widely accessible to beginning and intermediate ERP researchers.

The book provides detailed, practical advice about how to design, conduct, and interpret ERP experiments, along with the reasons why things should be done in a particular way. I did not attempt to provide comprehensive coverage of every possible way of recording and analyzing ERPs, because that would be too much for a beginning or intermediate researcher to digest. Instead, I've tried to provide a detailed treatment of the most basic techniques. I also tried to make the book useful for researchers who do not plan to conduct their own ERP studies, but who want to be able to understand and evaluate published or submitted ERP experiments. The book is aimed at cognitive neuroscientists, but it should also be useful for researchers in related fields, such as affective neuroscience and experimental psychopathology.

Acknowledgments

Several people have had a direct or indirect impact on the development of this book, and I would like to thank them for their help.

First, I'd like to thank Martha Neuringer of the Oregon Regional Primate Research Center and Dell Rhodes of Reed College, who provided my first introduction to ERPs.

Next, I'd like to thank Steve Hillyard and everyone who worked with me in the Hillyard lab at UCSD. Almost everything in this book can be attributed to the amazing group of people who worked in that lab in the late 1980s and early 1990s. In particular, I'd like to acknowledge Jon Hansen, Marty Woldorff, Ron Mangun, Marta Kutas, Cyma Van Petten, Steve Hackley, Hajo Heinze, Vince Clark, Paul Johnston, and Lourdes Anllo-Vento. I'd also like to acknowledge the inspiration provided by Bob Galambos, who taught me, among other things, that "you have to get yourself a phenomenon" at the beginning of the research enterprise.

I'd also like to acknowledge the important contributions of my first group of graduate students here at the University of Iowa, particularly Massimo Girelli, Ed Vogel, and Geoff Woodman. Many of the ideas in this book became crystallized as I taught them about ERPs, and they helped refine the ideas as we put them into practice. They will forever be my A-Team. I've also received many excellent comments and suggestions from my current students, Joo-seok Hyun, Weiwei Zhang, Jeff Johnson, Po-Han Lin, and Adam Niese. They suggested a number of additions that new ERP researchers will find very useful; they also tolerated my absence from the lab every morning for several months while I completed the book. My collaborator Max Hopf helped with chapter 7 by filling some gaps in my knowledge of MEG and source localization techniques.

I would also like to mention the generous financial support that has allowed me to pursue ERP research and write this book. In particular, the McDonnell-Pew Program in Cognitive Neuroscience supported my first research here at Iowa, as well as much of my graduate training, and I've also received generous support from the National Institute of Mental Health and the National Science Foundation. The University of Iowa has provided exceptional financial and administrative support, including a leave several years ago during which I started this book and a subsequent leave during which I was able to complete it. And above all, the James McKeen Cattell Fund provided a sabbatical award that was instrumental in providing me with the time and motivation to complete the book.

Finally, I would like to thank my family for providing emotional support and pleasant diversions as I worked on this book. Lisa helped me carve out the time that was needed to finish the book; Alison kept me from getting too serious; and Carter made sure that I was awake by 5:30 every morning so that I could get an early start on writing (and he could watch the Teletubbies).

1 An Introduction to Event-Related Potentials and Their Neural Origins

This chapter introduces the event-related potential (ERP) technique. The first section describes the goals of this book and discusses the perspective from which I've written it. The second section provides a brief history of the ERP technique. The third section describes two simple ERP experiments as examples that introduce some of the basic concepts of ERP experimentation. The fourth section describes the advantages and disadvantages of the ERP technique in relation to other techniques. The fifth section describes the neural and biophysical origins of ERPs and the associated event-related magnetic fields. The final section contains a brief description of the most commonly observed ERP components in cognitive neuroscience experiments.

Goals and Perspective

This book is intended as a guidebook for people who wish to use ERPs to answer questions of broad interest in cognitive neuroscience and related fields. This includes cognitive scientists who plan to use ERPs to address questions that are essentially about cognition rather than questions that are essentially about neuroscience. The book should also be very useful for researchers in the growing area of affective neuroscience, as well as those in the area of psychopathology. It also provides a good background for researchers and students who encounter ERP studies in the literature and want to be able to understand and evaluate them.

The book was written for people who are just starting to do ERP research and for people who have been doing it for a few years and would like to understand more about why things are done in a particular way. ERP experts may find it useful as a reference (and they would probably learn something new by reading chapter 5, which provides a fairly detailed account of filtering that is approachable for people who don't happen to have an advanced degree in electrical engineering).

The book provides practical descriptions of straightforward methods for recording and analyzing ERPs, along with the theoretical background to explain why these are particularly good methods. The book also provides some advice about how to design ERP experiments so that they will be truly useful in answering broadly significant questions (i.e., questions that are important to people who don't themselves conduct ERP experiments). Because the goal of this book is to provide an introduction to ERPs for people who are not already experts, I have focused on the most basic techniques and neglected many of the more sophisticated approaches (although I have tried to at least mention the most important of them). For a broader treatment, aimed at experts, see David Regan's massive treatise (Regan, 1989).

To keep things simple, this book focuses primarily on the techniques used in my own laboratory (and in many of the world's leading ERP labs). In most cases, these are techniques that I learned as a graduate student in Steve Hillyard's laboratory at University of California, San Diego, and they reflect a long history of electrophysiological recordings dating back to Hallowell Davis's lab in the 1930s (Davis was the mentor of Bob Galambos, who was the mentor of Steve Hillyard; Galambos was actually a subject in the first sensory ERP experiments, described in the next section). Other approaches to ERP experimentation may be just as good or even better, but the techniques described here have stood the test of time and provide an excellent foundation for more advanced approaches.

This book reflects my own somewhat idiosyncratic perspective on the use of ERP recordings in cognitive neuroscience, and there are two aspects of this perspective that deserve some comment. First, although much of my own research uses ERP recordings, I believe that the ERP technique is well suited to answering only a small subset of the questions that are important to cognitive neuroscientists. The key, of course, is figuring out which issues this technique best addresses. Second, I take a relatively low-tech approach to ERPs. In the vast majority of cases, I believe that it is better to use a modest number of electrodes and fairly simple data analysis techniques instead of a large array of electrodes and complicated data analysis techniques. This is heresy to many ERP researchers, but the plain fact is that ERPs are not a functional neuroimaging technique and cannot be used to definitively localize brain activity (except under a very narrow set of conditions). I also believe that too much has been made of brain localization, with many researchers seeming to assume that knowing *where* a cognitive process happens is the same as knowing *how* it happens. In other words, there is much more to cognitive neuroscience than functional neuroanatomy, and ERPs can be very useful in elucidating cognitive mechanisms and their neural substrates even when we don't know where the ERPs are generated.

A Bit of History

In 1929, Hans Berger reported a remarkable and controversial set of experiments in which he showed that one could measure the electrical activity of the human brain by placing an electrode on the scalp, amplifying the signal, and plotting the changes in voltage over time (Berger, 1929). This electrical activity is called the electroencephalogram, or EEG. The neurophysiologists of the day were preoccupied with action potentials, and many of them initially believed that the relatively slow and rhythmic brain waves Berger observed were some sort of artifact. After a few years, however, the respected physiologist Adrian (Adrian & Matthews, 1934) also observed human EEG activity, and Jasper and Carmichael (1935) and Gibbs, Davis, and Lennox (1935) confirmed the details of Berger's observations. These findings led to the acceptance of the EEG as a real phenomenon.

Over the ensuing decades, the EEG proved to be very useful in both scientific and clinical applications. In its raw form, however, the EEG is a very coarse measure of brain activity, and it is very difficult to use it to assess the highly specific neural processes that are the focus of cognitive neuroscience. The drawback of the EEG is that it represents a mixed up conglomeration of hundreds of different neural sources of activity, making it difficult to isolate individual neuro-cognitive processes. However, embedded within the EEG are the neural responses associated with specific sensory, cognitive, and motor events, and it is possible to extract these responses from the overall EEG by means of a simple averaging technique (and more sophisticated techniques, as well). These specific responses are called event-related potentials to denote the fact that they are electrical potentials associated with specific events.

As far as I can tell, the first unambiguous sensory ERP recordings from awake humans were performed in 1935–1936 by Pauline and Hallowell Davis, and published a few years later (Davis et al., 1939; Davis, 1939). This was long before computers were available for recording the EEG, but the researchers were able to see clear ERPs on single trials during periods in which the EEG was quiescent (the first published computer-averaged ERP waveform were apparently published by Galambos and Sheatz in 1962). Not much ERP work was done in the 1940s due to World War II, but research picked up again in the 1950s. Most of this research focused on sensory issues, but some of it addressed the effects of top-down factors on sensory responses.

The modern era of ERP research began in 1964, when Grey Walter and his colleagues reported the first cognitive ERP component, which they called the *contingent negative variation* or CNV (Walter et al., 1964). On each trial of this study, subjects were presented with a warning signal (e.g., a click) followed 500 or 1,000 ms later by a target stimulus (e.g., a series of flashes). In the absence of a task, each of these two stimuli elicited the sort of sensory ERP response that one would expect for these stimuli. However, if subjects were required to press a button upon detecting the target, a large negative voltage was observed at frontal electrode sites during the period that separated the warning signal and the target. This negative voltage—the CNV—was clearly not just a sensory response. Instead, it appeared to reflect the subject's preparation for the upcoming target. This exciting new finding led many researchers to begin exploring cognitive ERP components.

The next major advance was the discovery of the P3 component by Sutton, Braren, Zubin, and John (1965). They found that when subjects could not predict whether the next stimulus would be auditory or visual, the stimulus elicited a large positive P3 component that peaked around 300 ms poststimulus; this component was much smaller when the modality of the stimulus was perfectly predictable. They described this result in terms of information theory, which was then a very hot topic in cognitive psychology, and their paper generated a huge amount of interest. To get a sense of the impact of this study, I ran a quick Medline search and found about sixteen hundred journal articles that refer to the P300 (or P3) component in the title or abstract. This search probably missed at least half of the articles that talk about the P300 component, so this is an impressive amount of research. In addition, the Sutton et al. (1965) paper has been cited almost eight hundred times. There is no doubt that many millions of dollars have been spent on P300 studies (not to mention the many marks, pounds, yen, etc.).

Over the ensuing fifteen years, a great deal of research focused on identifying various cognitive ERP components and developing methods for recording and analyzing ERPs in cognitive experiments. Because people were so excited about being able to record human brain activity related to cognition, ERP papers in this period were regularly published in *Science* and *Nature* (much like the early days of PET and fMRI research). Most of this research was focused on discovering and understanding ERP components rather than using them to address questions of broad scientific interest. I like to call this sort of experimentation *ERPology*, because it is simply the study of ERPs.

ERPology plays an important role in cognitive neuroscience, because it is necessary to know quite a bit about specific ERP components before one can use them to study issues of broader importance. Indeed, a great deal of ERPology continues today, resulting in a refinement of our understanding of the components discovered in previous decades and the discovery of additional components. However, so much of ERP research in the 1970s was focused on ERPology that the ERP technique began to have a bad reputation among many cognitive psychologists and neuroscientists in the late 1970s and early 1980s. As time progressed, however, an increasing proportion of ERP research was focused on answering questions of broad scientific interest, and the reputation of the ERP technique began to improve. ERP research started becoming even more popular in the mid 1980s, due in part to the introduction of inexpensive computers and in part to the general explosion of research in cognitive neuroscience. When PET and the fMRI were developed, many ERP researchers thought that ERP research might die away, but exactly the opposite happened: because ERPs have a high temporal resolution that hemodynamic measures lack, most cognitive neuroscientists view the ERP technique as an important complement to PET and fMRI, and ERP research has flourished rather than withered.

Now that I've provided a brief history of the ERP technique, I'd like to clarify some terminology. ERPs were originally called *evoked potentials* (EPs) because they were electrical *potentials* that were *evoked* by stimuli (as opposed to the spontaneous EEG rhythms). The earliest published use of the term "event-related potential" that I could find was by Herb Vaughan, who in a 1969 chapter wrote,

Since cerebral processes may be related to voluntary movement and to relatively stimulus-independent psychological processes (e.g. Sutton et al., 1967; Ritter et al., 1968), the term "evoked potentials" is no longer sufficiently general to apply to all EEG phenomena related to sensorymotor processes. Moreover, suffi-

ciently prominent or distinctive psychological events may serve as time references for averaging, in addition to stimuli and motor responses. The term "event related potentials" (ERP) is proposed to designate the general class of potentials that display stable time relationships to a definable reference event. (Vaughan, 1969, p. 46)

Most research in cognitive neuroscience now uses the term *event-related potential*, but you might occasionally encounter other terms, especially in other fields. Here are a few common ones:

Evoked response. This means the same thing as *evoked potential*. *Brainstem evoked response (BER).* These are small ERPs elicited within the first 10 ms of stimulus onset by auditory stimuli such as clicks. They are frequently used in clinical audiology. They are also called *auditory brainstem responses* (ABRs) or *brainstem auditory evoked responses* (BAERs).
Visual evoked potential (VEP). This term is commonly used in clinical contexts to describe ERPs elicited by visual stimuli that are used to assess pathology in the visual system, such as demyelination caused by multiple sclerosis. A variant on this term is *visual evoked response* (VER).
Evoked response potential (ERP). This is apparently an accidental miscombination of evoked response and event-related potential (analogous to combining *irrespective* and *regardless* into *irregardless*).

A Simple Example Experiment

This section introduces the basics of the ERP technique. Rather than beginning with an abstract description, I will start by describing a simple ERP experiment that my lab conducted several years ago. This experiment was a variant on the classic *oddball* paradigm (which is really the same thing as the *continuous performance task* that is widely used in psychopathology research). Subjects viewed sequences consisting of 80 percent Xs and 20 percent Os, and they

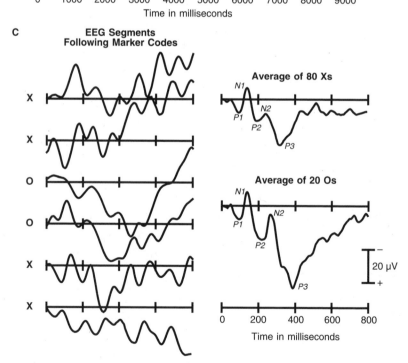

A

X

Stimulation Computer — Marker Codes

EEG

Filters & Amplifier → Digitization Computer

B

EEG Recorded from the Pz Electrode Site

X   X   O   O   X   X

20 μV
−
+

0   1000   2000   3000   4000   5000   6000   7000   8000   9000
Time in milliseconds

C

EEG Segments Following Marker Codes

X

X

O

O

X

X

Average of 80 Xs

N1
P1   N2
P2
P3

Average of 20 Os

N1
P1   N2
P2
P3

20 μV
−
+

0   200   400   600   800
Time in milliseconds

pressed one button for the Xs and another button for the Os. Each letter was presented on a video monitor for 100 ms, followed by a 1,400-ms blank interstimulus interval. While the subject performed this task, we recorded the EEG from several electrodes embedded in an electrode cap. The EEG was amplified by a factor of 20,000 and then converted into digital form for storage on a hard drive. Whenever a stimulus was presented, the stimulation computer sent marker codes to the EEG digitization computer, which stored them along with the EEG data (see figure 1.1A).

During a recording session, we viewed the EEG on the digitization computer, but the stimulus-elicited ERP responses were too small to discern within the much larger EEG. Figure 1.1B shows the EEG that was recorded at one electrode site (Pz, on the midline over the parietal lobes) from one of the subjects over a period of nine seconds. If you look closely, you can see that there is some consistency in the response to each stimulus, but it is difficult to see exactly what the responses look like. Note that negative is plotted upward in this figure (see box 1.1 for a discussion of this odd convention).

At the end of each session, we performed a simple signal-averaging procedure to extract the ERPs elicited by the Xs and the Os (see figure 1.1C). Specifically, we extracted the segment of EEG surrounding each X and each O and lined up these EEG segments with respect to the marker codes (which occurred at the onset of each stimulus). We then simply averaged together the single-trial waveforms, creating averaged ERP waveforms for the X and the O at each electrode site. For example, we computed the voltage at 24 ms poststimulus in the averaged X waveform by taking the voltage

◄ **Figure 1.1**   Example ERP experiment. The subject views frequent Xs and infrequent Os presented on a computer monitor while the EEG is recorded from a midline parietal electrode site. This signal is filtered and amplified, making it possible to observe the EEG. The rectangles show an 800-ms time epoch following each stimulus in the EEG. There is a great deal of trial-to-trial variability in the EEG, but a clear P3 wave can be seen following the infrequent O stimuli. The bottom of the figure shows averaged ERPs for the Xs and Os. Note that negative is plotted upward in this figure.

Box 1.1   Which Way Is Up?

It is a common, although not universal, convention to plot ERP waveforms with negative voltages upward and positive voltages downward. The sole reason that I plot negative upward is that this was how things were done when I joined Steve Hillyard's lab at UCSD. I once asked Steve Hillyard's mentor, Bob Galambos, how this convention came about. His answer was that that was simply that this was how things were done when he joined Hal Davis's lab at Harvard in the 1930s (see, e.g., Davis et al., 1939; Davis, 1939). Apparently, this was a common convention for the early physiologists. Manny Donchin told me that the early neurophysiologists plotted negative upward, possibly because this allows an action potential to be plotted as an upward-going spike, and this influenced manufacturers of early EEG equipment, such as Grass. Bob Galambos also mentioned that an attempt to get everyone to agree to a uniform positive-up convention was made in the late 1960s or early 1970s, but one prominent researcher (who will remain nameless) refused to switch from negative-up to positive-up, and the whole attempt failed.

At present, some investigators plot negative upward and others plot positive upward. This is sometimes a source of confusion, especially for people who do not regularly view ERP waveforms, and it would probably be a good idea for everyone to use the same convention (and we should probably use the same positive-up convention as the rest of the scientific world). Until this happens, plots of ERP waveforms should always make it clear which way is up.

ERP researchers should not feel too bad about this mixed up polarity problem. After all, fMRI researchers often reverse left and right when presenting brain images (due to a convention in neurology).

measured 24 ms after each X stimulus and averaging all of these voltages together. By doing this averaging at each time point following the stimulus, we end up with a highly replicable waveform for each stimulus type.

The resulting averaged ERP waveforms consist of a sequence of positive and negative voltage deflections, which are called *peaks*, *waves*, or *components*. In figure 1.1C, the peaks are labeled *P1*, *N1*, *P2*, *N2*, and *P3*. *P* and *N* are traditionally used to indicate positive-going and negative-going peaks, respectively, and the number simply indicates a peak's position within the waveform (it

Box 1.2   Component Naming Conventions

I much prefer to use names such as *N1* and *P3* rather than *N100* and *P300*, because a component's latency may vary considerably across experiments, across conditions within an experiment, or even across electrode sites within a condition. This is particularly true of the P3 wave, which almost always peaks well after 300 ms (the P3 wave had a peak latency of around 300 ms in the very first P3 experiment, and the name *P300* has persisted despite the wide range of latencies). Moreover, in language experiments, the P3 wave generally follows the N400 wave, making the term *P300* especially problematic. Consequently, I prefer to use a component's ordinal position in the waveform rather than its latency when naming it. Fortunately, the latency in milliseconds is often approximately 100 times the ordinal position, so that P1 = P100, N2 = N200, and P3 = P300. The one obvious exception to this is the N400 component, which is often the second major negative component. For this reason, I can't seem to avoid using the time-based name *N400*.

is also common to give a precise latency, such as P225 for a positive peak at 225 ms). Box 1.2 further discusses the labeling conventions for ERP components. The sequence of ERP peaks reflects the flow of information through the brain.

The initial peak (P1) is an obligatory sensory response that is elicited by visual stimuli no matter what task the subject is doing (task variations may influence P1 amplitude, but no particular task is necessary to elicit a P1 wave). In contrast, the P1 wave is strongly influenced by stimulus parameters, such as luminance. The early sensory responses are called *exogenous* components to indicate their dependence on external rather than internal factors. The P3 wave, in contrast, depends entirely on the task performed by the subject and is not directly influenced by the physical properties of the eliciting stimulus. The P3 wave is therefore termed an *endogenous* component to indicate its dependence on internal rather than external factors.

In the experiment shown in figure 1.1, the infrequent O stimuli elicited a much larger P3 wave than the frequent X stimuli. This is exactly what thousands of previous oddball experiments have found. If you're just beginning to get involved in ERP research, I would recommend running an oddball experiment like this as your first experiment. It's simple to do, and you can compare your results with a huge number of published experiments.

We conducted the averaging process separately for each electrode site, yielding a separate averaged ERP waveform for each combination of stimulus type and electrode site. The P3 wave shown in figure 1.1C was largest at a midline parietal electrode site but could be seen all over the scalp. The P1 wave, in contrast, was largest at lateral occipital electrode sites, and was absent over prefrontal cortex. Each ERP component has a distinctive scalp distribution that reflects the location of the patch of cortex in which it was originally generated. As I will discuss later in this chapter and in chapter 7, it is difficult to determine the location of the neural generator source simply by examining the distribution of voltage over the scalp.

Conducting an experiment like this has several steps. First, it is necessary to attach some sort of electrodes to the subject's scalp to pick up the EEG. The EEG must be filtered and amplified so that it can be stored as a set of discrete voltage measurements on a computer. Various artifacts (e.g., eyeblinks) may contaminate the EEG, and this problem can be addressed by identifying and removing trials with artifacts or by subtracting an estimate of the artifactual activity from the EEG. Once artifacts have been eliminated, averaging of some sort is usually necessary to extract the ERPs from the overall EEG. Various signal processing techniques (e.g., digital filters) are then applied to the data to remove noise[1] and isolate specific ERP components. The size and timing of the ERP components are then measured, and these measures are subjected to statistical analyses. The following chapters will cover these technical issues in detail. First, however, I'd like to provide an example of an experiment that was designed to answer a question of broad interest that could, in principle, be addressed by other methods but which was well suited for ERPs.

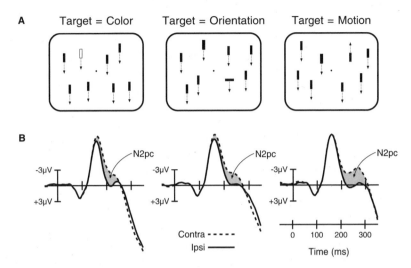

Figure 1.2 Example stimuli and data from Girelli and Luck's (1997) study. Three types of feature pop-outs were used: color, orientation, and motion. The waveforms show the averages from lateral occipital electrodes contralateral versus ipsilateral to the location of the pop-out. An N2pc component (indicated by the shaded region) can be seen as a more negative-going response between approximately 175 and 275 ms poststimulus. Negative is plotted upward. (Reprinted with permission from Luck and Girelli, 1998. © 1998 MIT Press.)

A Real Experiment

Figure 1.2 illustrates an experiment that Massimo Girelli conducted in my laboratory as a part of his dissertation (for details, see Girelli & Luck, 1997). The goal of this experiment was to determine whether the same attention systems are used to detect visual search targets defined by color and by motion. As is widely known, color and motion are processed largely independently at a variety of stages within the visual system, and it therefore seemed plausible that the detection of a color-defined target would involve different attention systems than the detection of a motion-defined target. However, there is also evidence that the same cortical areas may process both the color and the motion of a discrete object, at least under some conditions, so it also seemed plausible that the same attention systems would be used to detect discrete visual search targets defined by color or by motion.

Figure 1.2A shows the stimuli that we used to address this issue. On each trial, an array was presented consisting of eight moving objects, most of which were green vertical "distractor" bars that moved downward. On 25 percent of the trials, all of the bars were distractors; on the remaining trials, one of the bars differed from the distractors in color (red), orientation (vertical), or direction of motion (upward). We called these different bars *pop-out* stimuli, because they appeared to "pop out" from the otherwise homogeneous stimulus arrays. At the beginning of each trial block, the subjects were instructed that one of the three pop-out stimuli would be the target for that run, and they were told to press one button when an array contained the target and to press a different button for nontarget arrays. For example, when the horizontal bar was the target, the subjects would press one button when the array contained a horizontal pop-out and a different button when the array contained a color pop-out, a motion pop-out, or no pop-out. The main question was whether subjects would use the same attention systems to detect each of the three types of pop-outs.

To answer this question, we needed a way to determine whether a given attention system was used for a particular type of trial. To accomplish this, we focused on an attention-related ERP component called the *N2pc* wave, which is typically observed for visual search arrays containing targets. The N2pc component is a negative-going deflection in the N2 latency range (200–300 ms poststimulus) that is primarily observed at posterior scalp sites contralateral to the position of the target item (*N2pc* is an abbreviation of *N2-posterior-contralateral*). Previous experiments had shown that this component reflects the focusing of attention onto a potential target item (Luck & Hillyard, 1994a, 1994b), and we sought to determine whether the attention system reflected by this component would be present for motion-defined targets as well as color- and orientation-defined targets (which had been studied previously). Thus, we assumed that if the same attention-related ERP component was present for all three types of pop-outs, then the same attention system must be present in all three cases.

One of the most vexing issues in ERP experiments is the problem of assessing which ERP component is influenced by a given experimental manipulation. That is, the voltage recorded at any given time point reflects the sum of many underlying ERP components that overlap in time (see chapter 2 for an extended discussion of this issue). If we simply examined the ERP waveforms elicited by the color, orientation, and motion targets in this experiment, they might all have an N2 peak, but this peak might not reflect the N2pc component and might instead reflect some other neural activity that is unrelated to the focusing of attention. Fortunately, it is possible to use a trick to isolate the N2pc component (which is why we focused on this particular component). Specifically, the N2pc component is larger at electrode sites contralateral to the location of the target compared to ipsilateral sites, whereas the vast majority of ERP components would be equally large at contralateral and ipsilateral sites for these stimuli (because the overall stimulus array is bilateral). Thus, we can isolate the N2pc component by examining the difference in amplitude between the waveforms recorded from contralateral and ipsilateral electrode sites.

Figure 1.2B shows the waveforms we recorded at lateral occipital electrode sites for the three types of pop-out targets, with separate waveforms for contralateral and ipsilateral recordings. Specifically, the contralateral waveform is the average of the left hemisphere electrode site for right visual field targets and the right hemisphere site for left visual field targets, and the ipsilateral waveform is the average of the left hemisphere electrode site for left visual field targets and the right hemisphere site for right visual field targets. The difference between the contralateral and ipsilateral waveforms is the N2pc wave (indicated by the shaded area).

The main finding of this experiment was that an N2pc component was present for all three types of pop-out targets. The N2pc was larger for motion pop-outs than for color or orientation pop-outs, which was consistent with our impression that the motion pop-outs seemed to attract attention more automatically than the color and orientation pop-outs. To further demonstrate that the N2pc recorded for the three types of pop-outs was actually the same ERP component in all three cases, we examined the scalp distribution of the N2pc effect (i.e., the variations in N2pc amplitude across the different electrode sites). The scalp distribution was highly similar for the three pop-out types, and we therefore concluded that subjects used the same attention system across pop-out dimensions.

This experiment illustrates three main points. First, the N2pc effects shown in figure 1.2B are only about 2–3 μV (microvolts, millionths of a volt) in size. These are tiny effects. But if we use appropriate methods to optimize the signal-to-noise ratio, we can see these tiny effects very clearly. One of the main goals of this book is to describe procedures for obtaining the best possible signal-to-noise ratio (see especially chapters 2–4).

A second important point is that this experiment uses ERP recordings as tool to address a question that is fundamentally methodology independent (i.e., it is not an ERPology experiment). Although the central question of this experiment was not ERP-specific, it was a question for which ERPs were particularly well suited. For example, it is not clear how one could use purely behavioral methods to determine whether the same attention systems are used for these different stimuli, because different systems could have similar effects on behavioral output. Similarly, functional neuroimaging techniques such as PET and fMRI could be used to address this issue, but they would not be as revealing as the ERP data because of their limited temporal resolution. For example, if a given brain area were found to be active for all three pop-out types, it would not be clear whether this reflected a relatively early attention effect (such as the N2pc wave) or some higher level decision process (analogous to the P3 wave that was observed for all three pop-out types). Although the ERP data from this experiment cannot indicate which cortical region was responsible for the N2pc wave, the fact that nearly identical scalp distributions were obtained for all three pop-out types indicates that the same cortical regions were involved, and the additional timing information provided by the ERP recordings provides further evidence that the same attention effect was present for all three pop-out types. Note also that, although it would be useful to know where the N2pc is generated, this information was not necessary for answering the main question of the experiment.

A third point this experiment illustrates is that the use of ERPs to answer cognitive neuroscience questions usually depends on previous ERPology experiments. If we did not already know that the N2pc wave is associated with the focusing of attention in visual search, we would not have been able to conclude that the same *attention* system was used for all three pop-out types. Consequently, the conclusions from this experiment are only as strong as the previous studies showing that the N2pc reflects the focusing of attention. Moreover, our conclusions are valid only if we have indeed isolated the same functional ERP component that was observed in previous N2pc experiments (which is likely in this experiment, given the N2pc's distinctive contralateral scalp distribution). The majority of ERP experiments face these limitations, but it is sometimes possible to design an experiment in a manner that does not require the identification of a specific ERP component. These are often the most conclusive ERP experiments, and chapter 2 describes several examples in detail.

Reliability of ERP Waveforms

Figure 1.2 shows what are called *grand average* ERP waveforms, which is the term ERP researchers use to refer to waveforms created by averaging together the averaged waveforms of the individual subjects. Almost all published ERP studies show grand averages, and individual-subject waveforms are presented only rarely (grand averages were less common in the early days of ERP research due to a lack of powerful computers). The use of grand averages masks the variability across subjects, which can be both a good thing (because the variability makes it difficult to see the similarities) and a bad thing (because the grand average may not accurately reflect the pattern of individual results). In either case, it is worth considering what single-subject ERP waveforms look like.

Figure 1.3 shows an example of single-subject waveforms from an N2pc experiment. The left column shows waveforms from a lateral occipital electrode site in five individual subjects (from a total set of nine subjects). As you can see, there is a tremendous amount of variability in these waveforms. Every subject has a P1 peak and an N1 peak, but the relative and absolute amplitudes of these peaks are quite different from subject to subject (compare, e.g., subjects 2 and 3). Moreover, not all of the subjects have a distinct P2 peak, and the overall voltage from 200–300 ms is positive for three subjects, near zero for one subject, and negative for one subject. This is quite typical of the variability that one sees in an ERP experiment (note that I didn't fish around for unusual examples—this is a random selection).

What are the causes of this variability? To illustrate one part of the answer to this question, the first three rows of the right side of figure 1.3 show the waveforms from a single subject who participated in three sessions of the same experiment. There is some variability from session to session, but this variability is very small compared to the variability from subject to subject. I don't know of any formal studies of the variability of the ERP waveform, but the pattern shown in figure 1.3—low within-subject variability and high between-subject variability—is consistent with my experience. A variety of factors may cause the within-subject variability, ranging from global state factors (e.g., number of hours of sleep the previous night) to shifts in task strategy. John Polich has published an interesting series of studies showing that the P3 wave is sensitive to a variety of global factors, such as time since the last meal, body temperature, and even the time of year (see review by Polich & Kok, 1995).

There are several potential causes of between-subject variability. One factor that probably plays a large role is the idiosyncratic fold-

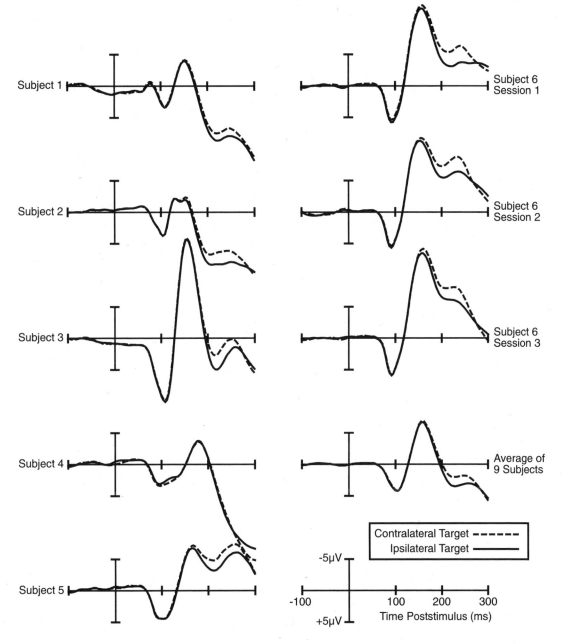

Figure 1.3 Example of the reliability of averaged ERP waveforms. Data from an N2pc experiment are shown for six individual subjects (selected at random from a total set of nine subjects). Subjects 1–5 participated in a single session, and subject 6 participated in three sessions. The lower right portion of the figure shows the average of all nine subjects from the experiment. Negative is plotted upward.

ing pattern of the cortex. As I will discuss later in this chapter, the location and orientation of the cortical generator source of an ERP component has a huge influence on the size of that component at a given scalp electrode site. Every individual has a unique pattern of cortical folding, and the relationship between functional areas and specific locations on a gyrus or in a sulcus may also vary. Although I've never seen a formal study of the relationship between cortical folding patterns and individual differences in ERP waveforms, I've always assumed that this is the most significant cause of waveform variation in healthy young adults (especially in the first 250 ms). There are certainly other factors that can influence the shape of the waveforms, including drugs, age, psychopathology, and even personality. But in experiments that focus on healthy young adults, these factors probably play a relative small role.

The waveforms in the bottom right portion of figure 1.3 represent the grand average of the nine subjects in this experiment. A striking attribute of the grand average waveforms is that the peaks are smaller than those in most of the single-subject waveforms. This might seem odd, but it is perfectly understandable. The time point at which the voltage reaches its peak values for one subject are not the same as for other subjects, and the peaks in the grand averages are not at the same time as the peaks for the individual subjects. Moreover, there are many time points at which the voltage is positive for some subjects and negative for others. Thus, the grand average is smaller overall than most of the individual-subject waveforms. This is a good example of how even simple data processing procedures can influence ERP waveforms in ways that may be unexpected.

There are studies showing that if you average together the pictures of a hundred randomly selected faces, the resulting average face is quite attractive and looks the same as any other average of a hundred randomly selected faces from the same population. In my experience, the same is true for ERP waveforms: Whenever I run an experiment with a given set of stimuli, the grand average of ten to fifteen subjects is quite attractive and looks a lot like the grand average of ten to fifteen different subjects in a similar experiment. Of course the waveform will look different if the stimuli or task differ substantially between experiments, and occasionally you will get several subjects with odd-looking waveforms that make the grand average look a little odd. But usually you will see a lot of similarity in the grand averages from experiment to experiment.

Advantages and Disadvantages of the ERP Technique

Comparison with Behavioral Measures

When ERPs were first used to study issues in the domain of cognitive neuroscience, they were primarily used as an alternative to measurements of the speed and accuracy of motor responses in paradigms with discrete stimuli and responses. In this context, ERPs have two distinct advantages. First, an overt response reflects the output of a large number of individual cognitive processes, and variations in reaction time (RT) and accuracy are difficult to attribute to variations in a specific cognitive process. ERPs, in contrast, provide a continuous measure of processing between a stimulus and a response, making it possible to determine which stage or stages of processing are affected by a specific experimental manipulation. As an example, consider the Stroop paradigm, in which subjects must name the color of the ink in which a word is drawn. Subjects are slower when the word is incompatible with the ink color than when the ink color and word are the same (e.g., subjects are slower to say "green" when presented with the word "red" drawn in green ink than when they are presented with the word "green" drawn in green ink). Do these slowed responses reflect a slowing of perceptual processes or a slowing of response processes? It is difficult to answer this question simply by looking at the behavioral responses, but studies of the P3 wave have been very useful in addressing this issue. Specifically, it is well documented that the latency of the P3 wave becomes longer when perceptual processes are delayed, but several studies have shown that P3 latency is not delayed on incompatible trials in the Stroop paradigm, indicating that the delays in RT reflect delays in some postperceptual stage (see, e.g., Duncan-Johnson & Kopell, 1981). Thus, ERPs are very useful for determining which stage or stages of processing are influenced by a given experimental manipulation (for a detailed set of examples, see Luck, Woodman, & Vogel, 2000).

A second advantage of ERPs over behavioral measures is that they can provide an online measure of the processing of stimuli even when there is no behavioral response. For example, much of my own research involves comparing the processing of attended versus ignored stimuli, and ERP recordings make it possible to monitor "covertly" the processing of the ignored stimuli without requiring subjects to respond to them. Similarly, ERP studies of language comprehension can assess the processing of a word embedded in the middle of a sentence at the time the word is presented rather than relying on a response made at the end of the sentence. Thus, the ability to covertly monitor the online processing of information is one of the greatest advantages of the ERP technique. For these reasons, I like to refer to the ERP technique as "reaction time for the twenty-first century."

ERP recordings also have some disadvantages compared to behavioral measures. The most obvious disadvantage is that the functional significance of an ERP component is virtually never as clear as the functional significance of a behavioral response. In most cases, we do not know the specific biophysical events that underlie the production of a given ERP response or the consequences of those events for information processing. In contrast, when a computer records a button-press response, we have a much clearer understanding of what that signal means. For example, when the reaction time (RT) in condition A is 30 ms longer than the RT in condition B, we know that the amount of time required to encode, process, and act on the stimuli was 30 ms longer in condition A than in condition B. In contrast, when the peak latency of an ERP component is 30 ms later in condition A than in condition B, we can draw no conclusions without relying on a long chain of assumptions and inferences (see chapter 2 for a discussion of some problems associated with measuring ERP latencies). Some amount of inference is always necessary when interpreting physiological measures of cognition, but some measures are easier to interpret than others. For example, when we record an action potential, we have an excellent understanding of the biophysical events that produced the action potential and the role that action potentials play in information processing. Thus, the basic signal is more difficult to interpret in ERP experiments than in behavioral experiments or in single-unit recordings (but probably no more difficult to interpret than the BOLD [blood oxygen level-dependent] signal in fMRI experiments).

A second disadvantage of the ERP technique is that ERPs are so small that it usually requires a large number of trials to measure them accurately. In most behavioral experiments, a reaction time difference can be observed with only about twenty to thirty trials per subject in each condition, whereas ERP effects often require fifty, a hundred, or even a thousand trials per subject in each condition. I have frequently started designing an ERP experiment and then given up when I realized that the experiment would require ten hours of data collection from each subject. Moreover, I rarely conduct an experiment that requires less than three hours of data collection from each subject. This places significant limitations on the types of questions that ERP recordings can realistically answer. For example, there are some behavioral experiments in which a given subject can receive only one trial in each condition (e.g., the inattentional blindness paradigm); experiments of this nature are not practical with ERP recordings.

Comparison with Other Physiological Measures

Table 1.1 compares the ERP technique (along with its magnetic counterpart, the event-related magnetic field, or ERMF, technique)

Table 1.1 Comparison of invasiveness, spatial resolution, temporal resolution, and cost for microelectrode measures (single-unit and local field-potential recordings), hemodynamic measures (PET and fMRI), and electromagnetic measures (ERPs and ERMFs)

	MICROELECTRODE MEASURES	HEMODYNAMIC MEASURES	ELECTROMAGNETIC MEASURES
Invasiveness	Poor	Good (PET) Excellent (fMRI)	Excellent
Spatial resolution	Excellent	Good	Undefined/poor (ERPs) Undefined/better (ERMFs)
Temporal resolution	Excellent	Poor	Excellent
Cost	Fairly expensive	Expensive (PET) Expensive (fMRI)	Inexpensive (ERPs) Expensive (ERMFs)

with several other physiological recording techniques along four major dimensions: invasiveness, spatial resolution, temporal resolution, and cost. The other classes of techniques considered are microelectrode measures (single-unit, multi-unit, and local field potential recordings) and hemodynamic measures (PET and fMRI).

Invasiveness Microelectrode measures require inserting an electrode into the brain and are therefore limited to nonhuman species (or, in rare cases, human neurosurgery patients). The obvious disadvantage of primate recordings is that human brains are different from primate brains. The less obvious disadvantage is that a monkey typically requires months of training to be able to perform a task that a human can learn in five minutes, and once a monkey is trained, it usually spends months performing the tasks while recordings are made. Thus, monkeys are often highly overtrained and probably perform tasks in a manner that is different from the prototypical naïve college sophomore. This can make it difficult to relate monkey results to the large corpus of human cognitive experiments. PET experiments are also somewhat problematic in terms of invasiveness. To avoid exposing subjects to excessive levels of radiation, each subject can be tested in only a small number of conditions. In contrast, there is no fundamental restriction on the amount of ERP or fMRI data that can be collected from a single subject.

Spatial and Temporal Resolution   Many authors have noted that electromagnetic measures and hemodynamic measures have complementary patterns of spatial and temporal resolution, with high temporal resolution and poor spatial resolution for electromagnetic measures and poor temporal resolution and high spatial resolution for hemodynamic measures. ERPs have a temporal resolution of 1 ms or better under optimal conditions, whereas hemodynamic measures are limited to a resolution of several seconds by the sluggish nature of the hemodynamic response. This is over a thousandfold difference, and it means that ERPs can easily address some questions that PET and fMRI cannot hope to address. However, hemodynamic measures have a spatial resolution in the millimeter range, which electromagnetic measures cannot match (except, perhaps, under certain unusual conditions). In fact, as I will discuss in greater detail later in this chapter and in chapter 7, the spatial resolution of the ERP technique is fundamentally undefined, because there are infinitely many internal ERP generator configurations that can explain a given pattern of ERP data. Unlike PET and fMRI, it is not currently possible to specify a margin of error for an ERP localization claim (for the typical case, in which several sources are simultaneously active). That is, with current techniques, it is impossible to know whether a given localization estimate is within some specific number of millimeters from the actual generator source. It may someday be possible to definitively localize ERPs, but at present the spatial resolution of the ERP technique is simply undefined.

The fact that ERPs are not easily localized has a consequence that is not often noted. Specifically, the voltage recorded at any given moment from a single electrode reflects the summed contributions from many different ERP generator sources, each of which reflects a different neurocognitive process. This makes it extremely difficult to isolate a single ERP component from the overall ERP waveform. This is probably the single greatest shortcoming of the ERP technique, because if you can't isolate an ERP component with confidence, it is usually difficult to draw strong conclusions. Chapter 2 will discuss this issue in greater detail.

Cost ERPs are much less expensive than the other techniques listed in table 1.1. It is possible to equip a good ERP lab for less than US $50,000, and the disposable supplies required to test a single subject are very inexpensive (US $1–3). A graduate student or an advanced undergraduate can easily carry out the actual recordings, and the costs related to storing and analyzing the data are minimal. These costs have dropped a great deal over the past twenty years, largely due to the decreased cost of computing equipment. FMRI is fairly expensive, the major costs being personnel and amortization of the machine. One session typically costs US $300–800. PET is exorbitantly expensive, primarily due to the need for radioactive isotopes with short half-lives and medical personnel. Single-unit recordings are also fairly expensive due to the per diem costs of maintaining the monkeys, the cost of the surgical and animal care facilities, and the high level of expertise required to record electrophysiological data from awake, behaving monkeys.

Choosing the Right Questions

Given that the ERP technique has both significant advantages and significant disadvantages, it is extremely important to focus ERP experiments on questions for which ERPs are well suited. For example, ERPs are particularly useful for addressing questions about which neurocognitive process is influenced by a given manipulation. Conversely, ERPs are poorly suited for asking questions that require neuroanatomical specificity (except under certain special conditions; see chapter 7). In addition, it is very helpful to ask questions that can be addressed with a component that is relatively easy to isolate (such as the N2pc component and the lateralized readiness potential) or questions that avoid the problem of identifying a specific ERP component altogether. Unfortunately, there are no simple rules for determining whether a given question can be easily answered with ERPs, but chapter 2 discusses a number of general principles.

The Neural Origins of ERPs

Basic Electrical Concepts

Before reading this section, I would encourage you to read the appendix, which reviews a few key principles of electricity and magnetism. You probably learned this material in a physics class several years ago, but it's worthwhile to review it again in the context of ERP recordings. You should definitely read it if (a) you're not quite sure what the difference between current and voltage is; or (b) you don't know Ohm's law off the top of your head. The appendix is, I admit, a bit boring. But it's short and not terribly complicated.

Electrical Activity in Neurons

To understand the nature of the voltages that can be recorded at the scalp, it is necessary to understand the voltages that are generated inside the brain. There are two main types of electrical activity associated with neurons, action potentials and postsynaptic potentials. Action potentials are discrete voltage spikes that travel from the beginning of the axon at the cell body to the axon terminals, where neurotransmitters are released. Postsynaptic potentials are the voltages that arise when the neurotransmitters bind to receptors on the membrane of the postsynaptic cell, causing ion channels to open or close and leading to a graded change in the potential across the cell membrane. If an electrode is lowered into the intercellular space in a living brain, both types of potentials can be recorded. It is fairly easy to isolate the action potentials arising from a single neuron by inserting a microelectrode into the brain, but it is virtually impossible to completely isolate a single neuron's postsynaptic potentials in an in vivo extracellular recording. Consequently, in vivo recordings of individual neurons ("single-unit" recordings) measure action potentials rather than postsynaptic potentials. When recording many neurons simultaneously, it is possible to measure either their summed postsynaptic potentials or their action potentials. Recordings of action potentials from large populations of neurons are called *multi-unit* recordings, and recordings of postsynaptic potentials from large groups of neurons are called *local field potential* recordings.

In the vast majority of cases, surface electrodes cannot detect action potentials due to the timing of the action potentials and the physical arrangement of axons. When an action potential is generated, current flows rapidly into and then out of the axon at one point along the axon, and then this same inflow and outflow occur at the next point along the axon, and so on until the action potential reaches a terminal. If two neurons send their action potentials down axons that run parallel to each other, and the action potentials occur at exactly the same time, then the voltages from the two neurons will summate and the voltage recorded from a nearby electrode will be approximately twice as large as the voltage recorded from a single action potential. However, if one neuron fires slightly after the other, then current at a given spatial location will be flowing into one axon at the same time that it is flowing out of the other axon, so they cancel each other and produce a much smaller signal at the nearby electrode. Because neurons rarely fire at precisely the same time (i.e., within microseconds of each other), action potentials in different axons will typically cancel, and the only way to record the action potentials from a large number of neurons is to place the electrode near the cell bodies and to use a very high impedance electrode that is sensitive only to nearby neurons. As a result, ERPs reflect postsynaptic potentials rather than action potentials (except under extremely rare circumstances).

Summation of Postsynaptic Potentials

Whereas the duration of an action potential is only about a millisecond, postsynaptic potentials typically last tens or even hundreds of milliseconds. In addition, postsynaptic potentials are largely confined to the dendrites and cell body and occur essentially instantaneously rather than traveling down the axon at a fixed rate. Under certain conditions, these factors allow postsynaptic potentials to summate rather than cancel, making it possible to record them at a great distance (i.e., at the scalp).

Very little research has examined the biophysical events that give rise to scalp ERPs, but figure 1.4 shows the current best guess. If an excitatory neurotransmitter is released at the apical dendrites of a cortical pyramidal cell, as shown in figure 1.4A, current will flow from the extracellular space into the cell, yielding a net negativity on the outside of the cell in the region of the apical dendrites. To complete the circuit, current will also flow out of the cell body and basal dendrites, yielding a net positivity in this area. Together, the negativity at the apical dendrites and the positivity at the cell body create a tiny *dipole* (a dipole is simply a pair of positive and negative electrical charges separated by a small distance).

The dipole from a single neuron is so small that it would be impossible to record it from a distant scalp electrode, but under certain conditions the dipoles from many neurons will summate, making it possible to measure the resulting voltage at the scalp. For the summated voltages to be recordable at the scalp, they must occur at approximately the same time across thousands or millions

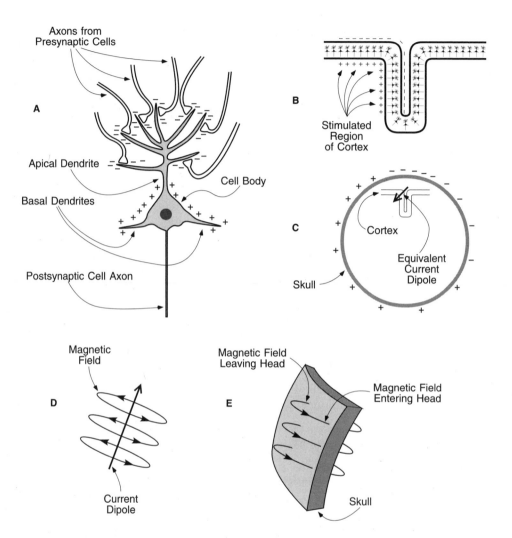

Figure 1.4   Principles of ERP generation. (A) Schematic pyramidal cell during neurotransmis-
sion. An excitatory neurotransmitter is released from the presynaptic terminals,
causing positive ions to flow into the postsynaptic neuron. This creates a net negative extracellular voltage (represented by the "−" symbols) in the area of other parts
of the neuron, yielding a small dipole. (B) Folded sheet of cortex containing many
pyramidal cells. When a region of this sheet is stimulated, the dipoles from the individual neurons summate. (C) The summated dipoles from the individual neurons
can be approximated by a single equivalent current dipole, shown here as an arrow.
The position and orientation of this dipole determine the distribution of positive and

of neurons, and the dipoles from the individual neurons must be spatially aligned. If the neurons are at random orientations with respect to each other, then the positivity from one neuron may be adjacent to the negativity from the next neuron, leading to cancellation. Similarly, if one neuron receives an excitatory neurotransmitter and another receives an inhibitory neurotransmitter, the dipoles of the neurons will be in opposite directions and will cancel. However, if the neurons all have a similar orientation and all receive the same type of input, their dipoles will summate and may be measurable at the scalp. This is most likely to occur in cortical pyramidal cells, which are aligned perpendicular to the surface of the cortex, as shown in figure 1.4B.

The summation of the individual dipoles is complicated by the fact that the cortex is not flat, but instead has many folds. Fortunately, however, physicists have demonstrated that the summation of many dipoles is essentially equivalent to a single dipole formed by averaging the orientations of the individual dipoles.[2] This averaged dipole is called an *equivalent current dipole* (ECD). It is important to note, however, that whenever the individual dipoles are more than 90 degrees from each other, they will cancel each other to some extent, with complete cancellation at 180 degrees. For example, the Purkinje cells in the cerebellar cortex are beautifully aligned with each other and oriented perpendicular to the cortical surface, but the cortical surface is so highly folded that the dipoles in one small patch of cerebellar cortex will almost always be cancelled by dipoles in a nearby but oppositely oriented patch, making it difficult or impossible to record cerebellar activity from the scalp.

Figure 1.4 (continued)
negative voltages recorded at the surface of the head. (D) Example of a current dipole with a magnetic field traveling around it. (E) Example of the magnetic field generated by a dipole that lies just inside the surface of the skull. If the dipole is roughly parallel to the surface, the magnetic field can be recorded as it leaves and enters the head; no field can be recorded if the dipole is oriented radially. Reprinted with permission from Luck and Girelli 1998. (© 1998 MIT Press.)

Volume Conduction

When a dipole is present in a conductive medium such as the brain, current is conducted throughout that medium until it reaches the surface. This is called *volume conduction* and is illustrated in figure 1.4C. The voltage that will be present at any given point on the surface of the scalp will depend on the position and orientation of the generator dipole and also on the resistance and shape of the various components of the head (most notably the brain, the skull, and the scalp; the eye holes also have an influence, especially for ERP activity generated in prefrontal cortex).

Electricity does not just run directly between the two poles of a dipole in a conductive medium, but instead spreads out through the conductor. Consequently, ERPs spread out as they travel through the brain. In addition, because electricity tends to follow the path of least resistance, ERPs tend to spread laterally when they encounter the high resistance of the skull. Together, these two factors greatly blur the surface distribution of voltage, and an ERP generated in one part of the brain can lead to substantial voltages at quite distant parts of the scalp. There are algorithms that can reduce this blurring, either by estimating the flow of current or by deblurring the voltage distribution to estimate the voltage distribution that is present on the brain's surface (Gevins et al., 1999; Pernier, Perrin, & Bertrand, 1988). These algorithms can be very useful, although you should remember that they only eliminate one source of blurring (the skull) and do not indicate the actual generator location of the ERPs.

Another important point is that electricity travels at nearly the speed of light. For all practical purposes, the voltages recorded at the scalp reflect what is happening in the brain at the same moment in time.

Magnetic Fields

The blurring of voltage caused by the high resistance of the skull can be largely circumvented by recording magnetic fields instead of electrical potentials. As figure 1.4D illustrates, an electrical dipole is always surrounded by a magnetic field, and these fields summate in the same manner as voltages. Thus, whenever an ERP is generated, a magnetic field is also generated, running around the ERP dipole. Moreover, the skull is transparent to magnetism,[3] and the magnetic fields are not blurred by the skull, leading to much greater spatial resolution than is possible with electrical potentials. The magnetic equivalent of the EEG is called the *magnetoencephalogram* (*MEG*), and the magnetic equivalent of an ERP is an *event-related magnetic field* (*ERMF*).

As figure 1.4E illustrates, a dipole that is parallel to the surface of the scalp will be accompanied by a magnetic field that leaves the head on one side of the dipole and enters back again on the other side. If you place a highly sensitive probe called a SQUID (super-conducting quantum interference device) next to the head, it is possible to measure the magnetic field as it leaves and reenters the head. Because magnetic fields are not as smeared out as electrical potentials, they can provide more precise localization. However, as chapter 7 will discuss, the combination of ERP and ERMF recordings provides even better localization than ERMF recordings alone. Unfortunately, magnetic recordings are very expensive because supercooling is expensive and because an expensive magnetically shielded recording chamber is necessary to attenuate the Earth's relatively large magnetic field.

ERP Localization

If I tell you the locations and orientations of a set of dipoles in a volume with a known distribution of conductances, then it would be possible for you to use a set of equations to compute the distribution of voltage that would be observed for those dipoles. This is called the *forward problem*, and it is relatively easy to solve. However, if I provide you with an observed voltage distribution and ask you to tell me the locations and orientations of the dipoles, you will not be able to provide an answer. This is called the *inverse problem*, and it is what mathematicians call an "ill-posed" or "underdetermined" problem. This simply means that there is not just one set of dipoles that can explain a given voltage distribution. In fact, researchers have known for over 150 years that an infinite number of different dipole configurations can produce any given voltage distribution (Helmholtz, 1853) (see also Nunez, 1981; Plonsey, 1963). Thus, it is impossible to know with certainty which one of these configurations is the one that is actually responsible for producing the observed voltage distribution.

It is possible to use additional constraints to overcome the ill-posed nature of the inverse problem. However, there is currently no mathematical technique that is widely accepted as providing a foolproof means of localizing ERP generator sources. By *foolproof*, I mean a technique that can provide a reasonably small and well-justified margin of error, making it possible to provide a p-value or likelihood ratio for a statement about the anatomical location of an ERP effect. For example, I would like to be able to say that the N2pc component in a particular experiment was generated within 12 mm of the fusiform gyrus and that the probability that this localization is incorrect is less than .05. I am aware of no mathematical localization technique that allows one to make such statements.

As discussed earlier in this chapter, the main advantages of the ERP technique are its high temporal resolution, its relatively low cost, its noninvasiveness, and its ability to provide a covert and continuous measure of processing. Spatial resolution is simply not one of the strengths of the ERP technique, and it seems appropriate to use this technique primarily to address issues that take advantage of its strengths and are not limited by its weaknesses.

Chapter 7 provides more details about ERP localization.

A Summary of Major ERP Components

This section provides a brief description of the ERP components commonly encountered in cognitive neuroscience research. Some of these components could justify an entire chapter, but I will cover just the basics. For more extensive descriptions, see reviews by Coles and Rugg (1995), Hillyard and Picton (1987), Picton and Stuss (1980), Näätänen and Picton (1986), and Regan (1989).

Before I start describing the components, however, I'd like to clarify something that is often confusing to ERP newcomers. Specifically, ERP components are usually given labels such as *P1* and *N1* that refer to their polarity and position within the waveform, and one must be careful not to assume that these labels are linked somehow to the nature of the underlying brain activity. Most notably, sensory components from different modalities that are given the same label are not usually related in any functional manner: They just happen to have the same polarity and ordinal position in the waveform. For example, the auditory P1 and N1 components bear no particular relationship to the visual P1 and N1 components. Some late components such as the P3 wave are largely modality-independent, but even the P3 wave may have modality-specific subcomponents (see, e.g., Luck & Hillyard, 1994a). Even within a single modality, a component labeled N2 in one experiment may not be the same as a component labeled N2 in another experiment. Some researchers use a bar over the latency of a component (e.g., P$\overline{300}$) when they are referring to a theoretical entity rather than simply labeling the observed polarity and latency of a peak from a particular experiment.

Visual Sensory Responses

C1   The first major visual ERP component is usually called the C1 wave, and it is largest at posterior midline electrode sites. Unlike most other components, it is not labeled with a P or an N because its polarity can vary. The C1 wave appears to be generated in area V1 (primary visual cortex), which in humans is folded into the calcarine fissure. The part of area V1 that codes the lower visual field is on the upper bank of the fissure and the part that codes the upper visual field is on the lower bank. As a result, the voltage recorded on the scalp above the calcarine fissure is positive for stimuli in the lower visual field and negative for stimuli in the upper visual field (Clark, Fan, & Hillyard, 1995; Jeffreys & Axford, 1972). The C1 wave is small or positive for stimuli on the horizontal midline, causing it to summate with the P1 wave into a single wave. Consequently, a distinct C1 wave is usually not observed unless upper-field stimuli are used to generate a negative C1 wave (which can be easily distinguished from the positive P1 wave). The C1 wave typically onsets 40–60 ms poststimulus and peaks 80–100 ms poststimulus, and it is highly sensitive to stimulus parameters, such as contrast and spatial frequency.

P1   The C1 wave is followed by the P1 wave, which is largest at lateral occipital electrode sites and typically onsets 60–90 ms poststimulus with a peak between 100–130 ms. Note, however, that P1 onset time is difficult to assess accurately due to overlap with the C1 wave. In addition, P1 latency will vary substantially depending on stimulus contrast. A few studies have attempted to localize the P1 wave by means of mathematical modeling procedures, sometimes combined with co-localization with fMRI effects, and these studies suggest that the early portion of the P1 wave arises from dorsal extrastriate cortex (in the middle occipital gyrus), whereas a later portion arises more ventrally from the fusiform gyrus (see Di Russo et al., 2002). Note, however, that at least thirty distinct visual areas are activated within the first 100 ms after the onset of a visual stimulus, and many of these areas presumably contribute to the voltages recorded in the C1 and P1 latency range. Like the C1 wave, the P1 wave is sensitive to a variations in stimulus parameters, as would be expected given its likely origins in extrastriate visual cortex. The P1 wave is also sensitive to the direction of spatial attention (see review by Hillyard, Vogel, & Luck, 1998) and to the subject's state of arousal (Vogel & Luck, 2000). Other top-down variables do not appear to reliably influence the P1 wave.

N1 The P1 wave is followed by the N1 wave. There are several visual N1 subcomponents. The earliest subcomponent peaks 100–150 ms poststimulus at anterior electrode sites, and there appear to be at least two posterior N1 components that typically peak 150–200 ms poststimulus, one arising from parietal cortex and another arising from lateral occipital cortex. Many studies have shown that spatial attention influences these components (see reviews by Hillyard et al., 1998; Mangun, 1995). In addition, the lateral occipital N1 subcomponent appears to be larger when subjects are performing discrimination tasks than when they are performing detection tasks, which has led to the proposal that this subcomponent reflects discriminative processing of some sort (Hopf et al., 2002; Ritter et al., 1979; Vogel & Luck, 2000).

P2 A distinct P2 wave follows the N1 wave at anterior and central scalp sites. This component is larger for stimuli containing target features, and this effect is enhanced when the targets are relatively infrequent (see Luck & Hillyard, 1994a). In this sense, the anterior P2 wave is similar to the P3 wave. However, the anterior P2 effects occur only when the target is defined by fairly simple stimulus features, whereas P3 effects can occur for arbitrarily complex target categories. At posterior sites, the P2 wave is often difficult to distinguish from the overlapping N1, N2, and P3 waves. Consequently, not much is known about the posterior P2 wave.

N170 and Vertex Positive Potential Jeffreys (1989) compared the responses to faces and non-face stimuli, and he found a difference between 150 and 200 ms at central midline sites that he named the *vertex positive potential* (the electrode site at the very top of the head is sometimes called the vertex site). Jeffreys noted that this effect inverted in polarity at more lateral sites, but he did not have any recordings from electrode sites over inferotemporal cortex. More recent studies from other laboratories that used a broader range of electrode sites have found that faces elicit a more negative potential than non-face stimuli at lateral occipital electrode sites, especially over the right hemisphere, with a peak at approximately 170 ms (Bentin et al., 1996; Rossion et al., 1999). This effect is typically called the *N170* wave. It is likely that the N170 and the vertex positive potential are just the opposite sides of the same dipole (although this is not a hundred percent certain—see George et al., 1996; Rossion et al., 1999).

The N170 is later and/or larger for inverted faces than for upright faces, a hallmark of face specialization. However, an inversion effect is also observed for non-face stimuli when the subjects have extensive experience viewing these stimuli in an upright orientations (Rossion et al., 2002). Moreover, other studies have shown that the vertex positive potential (and presumably the N170) also occurs for other highly familiar stimuli, such as words (Schendan, Ganis, & Kutas, 1998). But the face specificity of the N170 is still a topic of considerable debate (Bentin & Carmel, 2002; Carmel & Bentin, 2002; Rossion, Curran, & Gauthier, 2002).

Auditory Sensory Responses

Very Early Components   Under appropriate conditions, it is possible to observe a sequence of ERP peaks within the first 10 ms of the onset of an auditory stimulus. Various sources of evidence indicate that these peaks arise from various stages along the brainstem auditory pathways, and these peaks are therefore called the brainstem evoked responses (BERs) or auditory brainstem responses (ABRs). BERs are extremely useful for assessing auditory pathology, especially in infants. When my children were born, they were both given BER screening tests, and it was gratifying to see that a variant of my main research technique is used for such an important clinical application. The BERs are followed by the midlatency components (defined as responses between 10 and 50 ms), which probably arise at least in part from the medial geniculate nucleus and the primary auditory cortex. Attention has its first reliable effects in the midlatency range, but I don't know of any other top-down variables that influence auditory activity in this time range. The midlatency components are followed by the auditory P1 wave (ca. 50 ms), which is typically largest at frontocentral electrode sites.

N1 Like the visual N1 wave, the auditory N1 wave has several distinct subcomponents (see review by Näätänen & Picton, 1987). These include (1) a frontocentral component that peaks around 75 ms and appears to be generated in the auditory cortex on the dorsal surface of the temporal lobes, (2) a vertex-maximum potential of unknown origin that peaks around 100 ms, and (3) a more laterally distributed component that peaks around 150 ms and appears to be generated in the superior temporal gyrus. Further fractionation of the auditory N1 wave is possible (see, e.g., Alcaini et al., 1994). The N1 wave is sensitive to attention. Although some attention effects in the N1 latency range reflect the addition of an endogenous component, attention can influence the N1 wave itself (or at least some N1 subcomponents) (Woldorff et al., 1993).

Mismatch Negativity The mismatch negativity (MMN) is observed when subjects are exposed to a repetitive train of identical stimuli with occasional mismatching stimuli (e.g., a sequence with many 800-Hz tones and occasional 1200-Hz tones). The mismatching stimuli elicit a negative-going wave that is largest at central midline scalp sites and typically peaks between 160 and 220 ms. Several other components are sensitive to mismatches if they are task-relevant, but the MMN is observed even if subjects are not using the stimulus stream for a task (e.g., if they are reading a book while the stimuli are being presented). However, the MMN can be eliminated for stimuli presented in one ear if the subjects focus attention very strongly on a competing sequence of stimuli in the other ear (Woldorff, Hackley, & Hillyard, 1991). The MMN is thought to reflect a fairly automatic process that compares incoming stimuli to a sensory memory trace of preceding stimuli.

Somatosensory, Olfactory, and Gustatory Responses

The vast majority of cognitive ERP experiments use auditory or visual stimuli, so I will provide only a brief mention of components from other modalities. The response to a somatosensory stimulus begins with one of the rare ERP components (sometimes called *N10*) that reflects action potentials rather than postsynaptic potentials, arising from the peripheral nerves. This is followed by a set of subcortical components (ca. 10–20 ms) and a set of short- and medium-latency cortical components (ca. 20–100 ms). An N1 wave is then observed at approximately 150 ms, followed by a P2 wave at approximately 200 ms (together, these two peaks are sometimes called the *vertex potential*).

It is difficult to record olfactory and gustatory ERP responses, largely because it is difficult to deliver precisely timed, sudden-onset stimuli in these modalities (which is necessary when computing averaged ERP waveforms). However, recent studies have shown that these potentials can be recorded when using appropriate stimulation devices (see, e.g., Ikui, 2002; Wada, 1999).

The N2 Family

The N2 time range has been well studied, and researchers have identified many clearly different components in this time range (see extensive discussions in Luck & Hillyard, 1994a; Näätänen & Picton, 1986). As Näätänen and Picton (1986) describe, a repetitive, nontarget stimulus will elicit an N2 deflection that can be thought of as the *basic N2* (although it doubtless contains several subcomponents). If other stimuli (often called *deviants*) are occasionally presented within a repetitive train, a larger amplitude is observed in the N2 latency range. If these deviants are task-irrelevant tones, this effect will consist of a mismatch negativity (visual mismatches do not seem to elicit exactly this sort of automatic mismatch response). If the deviants are task-relevant, then a somewhat later N2 effect is also observed, called *N2b* (the mismatch negativity is sometimes called *N2a*). This component is larger for less frequent targets, and it is thought to be a sign of the stimulus categorization process. Both auditory and visual deviants will, if task-relevant, elicit an N2b component, but this effect is largest over central sites for auditory stimuli and over posterior sites for visual stimuli (Simson, Vaughan, & Ritter, 1977). We do not really know if the auditory and visual N2b components represent homologous neural processing functions.

In the visual domain, deviance is often studied spatially rather than temporally. That is, rather than examining the response to a deviant item presented within a temporal sequence of homogeneous items, one can compare the ERP waveform elicited by a simultaneous array of homogeneous items to the ERP waveform elicited by a simultaneous array that contains several identical items plus one deviant item. When this is done, one can distinguish three N2 components (Luck & Hillyard, 1994a). The first is a bilateral, anterior response that is present even when the deviant item is not a target (but it is not as automatic as the MMN because it is not present unless subjects are looking for deviant targets of some sort). This is followed by two posterior N2 subcomponents that are present only if the deviant item is a target (or resembles the target at first glance). One of these subcomponents is the standard N2b wave, which is bilateral and probability sensitive. The second is called *N2pc*, where the *pc* is an abbreviation of *posterior contralateral*, denoting the fact that this component is observed at posterior electrode sites contralateral to the location of the target. The N2pc component is not probability sensitive, and it reflects the focusing of spatial attention onto the target location (and possibly the suppression of the surrounding nontarget items—see Eimer, 1996; Luck et al., 1997; Luck & Hillyard, 1994b). A contralateral negativity is also observed during visual working memory tasks, but it has a more parietally focused scalp distribution and appears to reflect some aspect of working memory maintenance (Vogel & Machizawa, 2004).

The P3 Family

There are several distinguishable ERP components in the time range of the P3 wave. Squires, Squires, and Hillyard (1975) made the first major distinction, identifying a frontally maximal P3a component and a parietally maximal P3b component. Both were elicited by unpredictable, infrequent shifts in tone pitch or intensity, but the P3b component was present only when these shifts were task-relevant. When ERP researchers (including myself) refer to the *P3 component* or the *P300 component*, they almost always mean the *P3b component* (in fact, I will simply use the term *P3* to refer to the P3b component for the rest of this book). Other studies have shown that an unexpected, unusual, or surprising task-irrelevant stimulus within an attended stimulus train will elicit a frontal P3-like response (e.g., Courchesne, Hillyard, & Galambos, 1975; Polich & Comerchero, 2003; Soltani & Knight, 2000), but it is not clear whether this response is related to the P3a component as Squires, Squires and Hillyard originally described (1975). For example, Verleger, Jaskowski, and Wauschkuhn (1994) provided evidence that the P3b component is observed for targets that are infrequent but are in some sense expected (or "awaited" in the terms of this paper), whereas the frontal P3 wave is elicited by stimuli that are truly unexpected or surprising. However, it is not clear that this frontal P3 is as automatic as the P3a Squires and colleagues (1975) observed.

Given the thousands of published P3 experiments, you might think that we would have a very thorough understanding of the P3 wave. But you'd be wrong! We know a great deal about the effects of various manipulations on P3 amplitude and latency, but there is no clear consensus about what neural or cognitive process the P3 wave reflects.[4] Donchin (1981) proposed that the P3 wave is somehow related to a process he called "context updating" (updating one's representation of the current environment), but this proposal was not followed by a convincing set of experiments providing direct support for it. This is probably due to the fact that we don't have a good theory of context updating that specifies how it varies according to multiple experimental manipulations. If you are interested in the P3 wave, you should probably read Donchin's original proposal (Donchin, 1981), Verleger's extensive critique of the proposal (Verleger, 1988), and the response of Donchin and Coles to this critique (Donchin & Coles, 1988). In my own laboratory's research on attention, we have frequently assumed that the context-updating proposal is at least approximately correct, and this has led to a variety of very sensible results (e.g., Luck, 1998b; Vogel & Luck, 2002; Vogel, Luck, & Shapiro, 1998). But this assumption certainly carries some risk, so you should be careful in making assumptions about the meaning of the P3 wave.

Although we do not know exactly what the P3 wave means, we do know what factors influence its amplitude and latency (for extensive reviews of the early P3 literature, see Johnson, 1986; Pritchard, 1981; for more recent reviews, see Picton, 1992; Polich, 2004; Polich & Kok, 1995). The hallmark of the P3 wave is its sensitivity to target probability: As Duncan-Johnson and Donchin (1977) described in excruciating detail, P3 amplitude gets larger as target probability gets smaller. However, it is not just the overall probability that matters; local probability also matters, because the P3 wave elicited by a target becomes larger when it has been preceded by more and more nontargets. Moreover, it is the probability of the task-defined stimulus class that matters, not the probability of the physical stimulus. For example, if subjects are asked to press a button when detecting male names embedded in a sequence containing male and female names, with each individual name occurring only once, the amplitude of the P3 wave will depend on the relative proportions of male and female names in the sequence (see Kutas, McCarthy, & Donchin, 1977). Similarly, if the target is the letter E, occurring on 10 percent of trials, and the nontargets are selected at random from the other letters of the alphabet, the target will elicit a very large P3 wave even though the target letter is approximately four times more probable than any individual nontarget letter (see Vogel, Luck, & Shapiro, 1998).

P3 amplitude is larger when subjects devote more effort to a task, leading to the proposal that P3 amplitude can be used as a measure of resource allocation (see, e.g., Isreal et al., 1980). However, P3 amplitude is smaller when the subject is uncertain of whether a given stimulus was a target or nontarget. Thus, if a task is made more difficult, this might increase P3 amplitude by encouraging subjects to devote more effort to the task, but it might decrease P3 amplitude by making subjects less certain of the category of a given stimulus. Johnson (1984, 1986) proposed that the variables of probability (P), uncertainty (U), and resource allocation (R) combine to influence P3 amplitude in the following manner: P3 amplitude $= U \times (P + R)$.

Because the P3 wave depends on the probability of the task-defined category of a stimulus, it is logically necessary that the P3 wave must be generated after the stimulus has been categorized according to the rules of the task. Consequently, any manipulation that postpones stimulus categorization (including increasing the time required for low-level sensory processing or higher-level categorization) must increase P3 latency. This is logical, and countless studies have confirmed this prediction. Although P3 latency must logically depend on the time required to categorize the stimulus, it is not logically dependent on post-categorization processes; several studies have shown that P3 latency is not sensitive to the amount of time required to select and execute a response once a stimulus has been categorized (see, e.g., Kutas, McCarthy, & Donchin, 1977; Magliero et al., 1984). For example, if subjects press a left-hand button when they see the stimulus LEFT and right-hand button when they see the stimulus RIGHT, P3 latency is no faster or slower than when they are asked to make a left-hand response for RIGHT and a right-hand response for LEFT (which is known to increase the time required to perform stimulus-response mapping). In contrast, if the stimuli are perceptually degraded, then P3 latency is delayed for these stimuli. Thus, one can use P3 latency to determine if a given experimental manipulation influences the processes leading up to stimulus categorization or processes related to response selection and execution (for an example, see Luck, 1998b; for contrasting viewpoints, see Leuthold & Sommer, 1998; Verleger, 1997).

Language-Related ERP Components

The best studied language-related component is the N400, first reported by Kutas and Hillyard (1980) (see also the more recent review by Kutas, 1997). The N400 is negative-going wave that is usually largest over central and parietal electrode sites, with a slightly larger amplitude over the right hemisphere than over the left hemisphere. The N400 is typically seen in response to violations of semantic expectancies. For example, if sentences are presented one word at a time on a video monitor, a large N400 will be elicited by the last word of the sentence, "While I was visiting my home town, I had lunch with several old shirts." Little N400 activity would be observed if the sentence had ended with "friends" rather than "shirts." An N400 can also be observed to the second word in a pair of words, with a large N400 elicited by "tire ... sugar" and a small N400 elicited by "flour ... sugar." Some N400 activity is presumably elicited by any content word you read or hear, and relatively infrequent words such as "monocle" elicit larger N400s than relatively frequent words such as "milk."

Nonlinguistic stimuli can also elicit an N400 (or N400-like activity), as long as they are meaningful. For example, a line drawing will elicit an N400 if it is inconsistent with the semantic context created by a preceding sequence of words or line drawings (Ganis, Kutas, & Sereno, 1996; Holcomb & McPherson, 1994). However, it is possible that subjects named the stimuli subvocally, so it is possible that the N400 component reflects language-specific brain activity.

Although typically larger at right-hemisphere electrodes than left-hemisphere electrodes, the N400 appears to be generated primarily in the left temporal lobe. One can explain this apparent discrepancy by assuming that the generator dipole near the base of the left hemisphere does not point straight upward, but instead points somewhat medially. Studies of split-brain patients and lesion patients have shown that the N400 depends on left-hemisphere activity (Hagoort, Brown, & Swaab, 1996; Kutas, Hillyard, & Gazzaniga, 1988), and recordings from the cortical surface in neurosurgery patients have found clear evidence of N400-like activity in the left anterior medial temporal lobe (e.g., McCarthy et al., 1995).

Syntactic violations also elicit distinctive ERP components. One of these is called *P600* (see Osterhout & Holcomb, 1992, 1995). For example, the word "to" elicits a larger P600 in the sentence "The broker persuaded to sell the stock" than in the sentence "The broker hoped to sell the stock." Syntactic violations can also elicit a left frontal negativity from approximately 300–500 ms, which may be the same effect observed when wh-questions (e.g., "What is the …") are compared to yes-no questions (e.g., "Is the …"). Given the important distinction between syntax and semantics, it should not be surprising that words that are primarily syntactic in nature elicit different ERP activity than words with rich semantics. In particular, function words (e.g., to, with) elicit a component called *N280* at left anterior electrode sites, and this component is absent for content words (e.g., nouns and verbs). In contrast, content words elicit an N400 that is absent for function words.

Error Detection

In most ERP studies, researchers simply throw out trials with incorrect behavioral responses. However, by comparing the ERP waveform elicited on error trials with the ERP waveform elicited on correct trials, it is possible to learn something about the cause of the error and the brain's response following detection of the error. For example, Gehring et al. (1993) had subjects perform a speeded response task in which they responded so fast that they occasionally made errors that were obvious right away ("Oops! I meant to press the left button!"). When they compared the ERPs on correct trials to the ERPs on error trials, they observed a negative-going deflection at frontal and central electrode sites beginning just after the time of the response. Gehring et al. called this deflection the *error-related negativity* (ERN); it was independently discovered by Falkenstein et al., (1990), who called it the N_e. This component is often followed by a positive deflection called P_e. More recent studies have demonstrated that the ERN can be elicited by negative feedback following an incorrect response (Gehring & Willoughby, 2002) or by observing someone else making an incorrect response (van Schie et al., 2004).

Most investigators believe that the ERN reflects the activity of a system that either monitors responses or is sensitive to conflict between intended and actual responses. Evidence from fMRI and single-unit recordings suggests that these functions occur in the anterior cingulate cortex (Holroyd et al., 2004; Ito et al., 2003), and a dipole source modeling study showed that the scalp distribution of the ERN is consistent with a generator source in the anterior cingulate (Dehaene, Posner, & Tucker, 1994). However, it is difficult to localize a broadly distributed component such as the ERN with much precision on the basis of the observed distribution of voltage. An intracranial recording study found evidence of an ERN-like response in the anterior cingulate, which is more convincing, but ERN-like responses were also observed at many other cortical sites in this study. The generator of the ERN is therefore not yet known with certainty.

Response-Related ERP Components

If subjects are instructed to make a series of occasional manual responses, with no eliciting stimulus, the responses are preceded by a slow negative shift at frontal and central electrode sites that begins up to one second before the actual response. This is called the *bereitschaftspotential* (BP) or *readiness potential* (RP), and it was independently discovered by Kornhuber and Deecke (1965) and Vaughn, Costa, and Ritter (1968). The scalp topography of the readiness potential depends on which effectors will be used to make the response, with differences between the two sides of the body and differences within a given side.

The lateralized portion of the readiness potential is called the *lateralized readiness potential* (LRP), and it has been widely used in cognitive studies. As discussed in chapter 2, the LRP is particularly useful because it can be easily isolated from other ERP components. That is, because it is lateralized with respect to the hand making the response, whereas other components are not lateralized, it is easy to tell when a given experimental manipulation has affected the time or amplitude of the LRP. In contrast, it is difficult to be certain that a given experimental manipulation has influenced the P3 component rather than some other overlapping component, and this is one of the main reasons why it has been so difficult to determine what cognitive process the P3 component reflects.

The LRP is generated, at least in part, in motor cortex (Coles, 1989; Miller, Riehle, & Requin, 1992). The most interesting consequence of this is that the LRP preceding a foot movement is opposite in polarity to the LRP preceding a hand movement, reflecting the fact that the motor cortex representation of the hand is on the lateral surface of the brain, whereas the representation of the foot is on the opposed mesial surface. The LRP appears to reflect some key aspect of response preparation: responses are faster when the LRP is larger at the moment of stimulus onset, and there is a threshold level of LRP amplitude beyond which a response will inexorably be triggered (Gratton et al., 1988).

The RP and LRP may be present for hundreds of milliseconds before the response, but other components are more tightly synchronized to the response. The early view was that a positive-going deflection is superimposed on the RP beginning 80–90 ms before the response, followed by a negative-going deflection during the response and another positive-going deflection after the response. However, more recent research has identified many more movement-related components and subcomponents (see, e.g., Nagamine et al., 1994; Shibasaki, 1982).

The contingent negative variation (CNV), described in the brief history at the beginning of this chapter, should be mentioned again here because it is partly related to motor preparation. As you will recall, the CNV is a broad negative deflection between a warning stimulus and a target stimulus (Walter et al., 1964). When the period between the warning and target stimuli is lengthened to several seconds, it is possible to see that the CNV actually consists of a negativity following the warning stimulus, a return to baseline, and then a negativity preceding the target stimulus (Loveless & Sanford, 1975; Rohrbaugh, Syndulko, & Lindsley, 1976). The first negative phase is usually regarded as reflecting processing of the warning stimulus, and the second negative phase is usually regarded as reflecting the readiness potential that occurs as the subject prepares to respond to the target.

Suggestions for Further Reading

The following is a list of journal articles, books, and book chapters that provide broad and insightful discussions of general ERP issues.

Coles, M. G. H. (1989). Modern mind-brain reading: Psychophysiology, physiology and cognition. *Psychophysiology*, *26*, 251–269.

Coles, M. G. H., Smid, H., Scheffers, M. K., & Otten, L. J. (1995). Mental chronometry and the study of human information processing. In M. D. Rugg & M. G. H. Coles (Eds.), *Electrophysiology of Mind: Event-Related Brain Potentials and Cognition.* (pp. 86–131). Oxford: Oxford University Press.

Donchin, E. (1979). Event-related brain potentials: A tool in the study of human information processing. In H. Begleiter (Ed.), *Evoked Brain Potentials and Behavior* (pp. 13–88). New York: Plenum Press.

Donchin, E. (1981). Surprise!... Surprise? *Psychophysiology*, *18*, 493–513.

Gaillard, A. W. K. (1988). Problems and paradigms in ERP research. *Biological Psychology, 26*, 91–109.

Hillyard, S. A., & Picton, T. W. (1987). Electrophysiology of cognition. In F. Plum (Ed.), *Handbook of Physiology: Section 1. The Nervous System: Volume 5. Higher Functions of the Brain, Part 2* (pp. 519–584). Bethesda, MD: Waverly Press.

Kutas, M., & Dale, A. (1997). Electrical and magnetic readings of mental functions. In M. D. Rugg (Ed.), *Cognitive Neuroscience. Studies in Cognition* (pp. 197–242). Cambridge, MA: MIT Press.

Lindsley, D. B. (1969). Average evoked potentials—achievements, failures and prospects. In E. Donchin & D. B. Lindsley (Eds.), *Average Evoked Potentials: Methods, Results and Evaluations* (pp. 1–43). Washington, D.C.: U.S. Government Printing Office.

Nunez, P. L. (1981). *Electric Fields of the Brain*. New York: Oxford University Press.

Picton, T. W., & Stuss, D. T. (1980). The component structure of the human event-related potentials. In H. H. Kornhuber & L. Deecke (Eds.), *Motivation, Motor and Sensory Processes of the Brain, Progress in Brain Research* (pp. 17–49). North-Holland: Elsevier.

Sutton, S. (1969). The specification of psychological variables in average evoked potential experiments. In E. Donchin & D. B. Lindsley (Eds.), *Averaged Evoked Potentials: Methods, Results and Evaluations* (pp. 237–262). Washington, D.C.: U.S. Government Printing Office.

Vaughan, H. G., Jr. (1969). The relationship of brain activity to scalp recordings of event-related potentials. In E. Donchin & D. B. Lindsley (Eds.), *Average Evoked Potentials: Methods, Results and Evaluations* (pp. 45–75). Washington, D.C.: U.S. Government Printing Office.

2 The Design and Interpretation of ERP Experiments

This chapter discusses some of the central issues in the design and interpretation of ERP experiments. Many such issues are unique to a given research topic (e.g., equating word frequencies in language experiments), but this chapter will focus on a set of principles that are common to most cognitive ERP studies.[1] Throughout the chapter, I will distill the most significant points into a set of rules and strategies for designing and analyzing ERP experiments.

It may seem odd to place a chapter on experimental design and interpretation before the chapters that cover basic issues such as electrodes and averaging. However, the experimental design is the most important element of an ERP experiment, and the principles of experimental design have implications for the more technical aspects of ERP research.

Waveform Peaks versus Latent ERP Components

The term *ERP component* refers to one of the most important and yet most nebulous concepts in ERP research. An ERP waveform unambiguously consists of a series of peaks and troughs, but these voltage deflections reflect the sum of several relatively independent underlying or *latent* components. It is extremely difficult to isolate the latent components so that they can be measured independently, and this is the single biggest roadblock to designing and interpreting ERP experiments. Consequently, one of the keys to successful ERP research is to distinguish between the observable peaks of the waveform and the unobservable latent components. This section describes several of the factors that make it difficult to assess the latent components, along with a set of "rules" for avoiding misinterpreting the relationship between the observable peaks and the underlying components.

Voltage Peaks Are not Special

Panels A–C in figure 2.1 illustrate the relationship between the visible ERP peaks and the latent ERP components. Panel A shows an ERP waveform, and panel B shows a set of three latent ERP components that, when summed together, equal the ERP waveform in panel A. When several voltages are present simultaneously in a conductor such as the brain, the combined effect of the individual voltages is exactly equal to their sum, so it is quite reasonable to think about ERP waveforms as an expression of several summed latent components. In most ERP experiments, the researchers want to know how an experimental manipulation influences a specific latent component, but we don't have direct access to the latent components and must therefore make inferences about them from the observed ERP waveforms. This is more difficult than it might seem, and the first step is to realize that the maximum and minimum voltages (i.e., the peaks) in an observed ERP waveform are not necessarily a good reflection of the latent components. For example, the peak latency of peak 1 in the ERP waveform in panel A is much earlier than the peak latency of component C1 in panel B. This leads to our first rule of ERP experimental design and interpretation:

Rule 1. Peaks and components are not the same thing. There is nothing special about the point at which the voltage reaches a local maximum.

In light of this fundamental rule, I am always amazed at how often researchers use peak amplitude and peak latency to measure the magnitude and timing of ERP components. These measures often provide a highly distorted view of the amplitude and timing of the latent components, and better techniques are available for quantifying ERP data (see chapter 5).

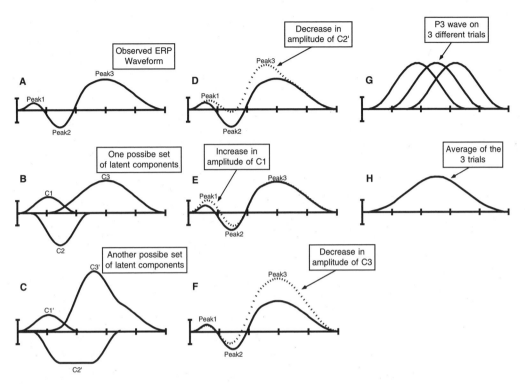

Figure 2.1 Examples of the latent components that may sum together to form an observed ERP waveform. Panels B and C show two different sets of latent components that could underlie the waveform shown in panel A. Panel D shows the effect of decreasing the amplitude of component C2′ by 50 percent (broken line) compared to the original waveform (solid line). Panel E shows how an increase in the amplitude of component C1 (broken line) relative to the original waveform (solid line) can create an apparent shift in the latencies of both peak 1 and peak 2. Panel F shows how an increase in the amplitude of component C3 (broken line) relative to the original waveform (solid line) can influence both the amplitude and the latency of peak 2. Panel G shows a component at three different latencies, representing trial-by-trial variations in timing; panel H shows the average of these three waveforms, which is broader and has a smaller peak amplitude (but the same area amplitude) compared to each of the single-trial waveforms.

Peak Shapes Are not the Same as Component Shapes

Panel C of figure 2.1 shows another set of latent components that also sum together to equal the ERP waveform shown in panel A. In this case, the relatively short duration and hill-like shape of peak 2 in panel A bears little resemblance to the long duration, boxcar-like component C2′ in panel C. This leads to our second rule:

Rule 2. It is impossible to estimate the time course or peak latency of a latent ERP component by looking at a single ERP waveform—there may be no obvious relationship between the shape of a local part of the waveform and the underlying components.

Violations of this rule are especially problematic when comparing two or more ERP waveforms. For example, consider the ERP waveforms in panel D of figure 2.1. The solid waveform represents the sum of the three latent components in panel C (which is the same as the ERP waveform in panel A), and the dashed waveform shows the effects of decreasing component C2′ by 50 percent. To make this a bit more concrete, you can think of these waveforms as the response to an attended stimulus and an unattended stimulus, respectively, such that ignoring the stimulus leads to a 50 percent decline in the amplitude of component C2′. Without knowing the underlying component structure, it would be tempting to conclude from the ERP waveforms shown in panel D that the unattended stimulus elicits not only a decrease in the amplitude of component C2′ but also an increase in the amplitude of component C1′ and a decrease in the latency and an increase in the amplitude of component C3′. In other words, researchers often interpret an effect that overlaps with multiple peaks in the ERP waveform as reflecting changes in multiple underlying components, but this interpretation is often incorrect. Alternatively, you might conclude from the waveforms in panel D that the attentional manipulation adds an additional, long-duration component that would not otherwise be present at all. This would also be an incorrect conclusion, which leads us to:

Rule 3. It is dangerous to compare an experimental effect (i.e., the difference between two ERP waveforms) with the raw ERP waveforms.

Rule 3 applies to comparisons of scalp distribution as well as comparisons of timing. That is, it is not usually appropriate to compare the scalp distribution of an experimental effect with the scalp distribution of the raw ERP waveform. For example, imagine you found that words elicited a more negative response in the N1 latency range than nonwords. To determine whether this effect reflects a change in the amplitude of the exogenous N1 or the addition of a word-specific neural process, it would be tempting to calculate a word-minus-nonword difference wave and compare the scalp distribution of this difference wave (in the N1 latency range) to the scalp distribution of the N1 elicited by nonwords. If the scalp distributions were found to be different, one might conclude that the effect represents the addition of endogenous, word-specific brain activity. But this conclusion would be unwarranted. The nonword-elicited N1 scalp distribution certainly represents overlapping activity from multiple latent components, and a different scalp distribution would be expected for the word-minus-nonword difference wave if a subset of these latent components were larger (or smaller) for words than for nonwords. Moreover, a change in the latency of one of the latent components would also lead to differences in scalp distribution. Thus, if the scalp distribution differs between two conditions, the only certain conclusion is that the experimental manipulation does not simply change the amplitude of all components proportionately.

This raises an important point about the relationship between amplitude and latency. Although the amplitude and latency of a latent component are conceptually independent, amplitude and latency often become confounded when measuring ERP waveforms. This occurs because the latent components overlap in time. Consider, for example, the relatively straightforward correspondence between the peaks in panel A of figure 2.1 and the latent components in panel B of the figure. Panel E of the figure shows the effects of increasing the amplitude of the first latent component on the summed ERP activity. When the amplitude of component A is increased by 50 percent, this creates an increase in the latency of both peak 1 and peak 2 in the summed waveform, and it also causes a decrease in the peak amplitude of peak 2. Panel F illustrates the effect of doubling the amplitude of the component C3, which causes a decrease in the amplitude and the latency of the second peak. Once again, this shows how the peak voltage in a given time range is a poor measure of the underlying ERP components in that latency range. This leads to our next rule:

Rule 4. Differences in peak amplitude do not necessarily correspond with differences in component size, and differences in peak latency do not necessarily correspond with changes in component timing.

Distortions Caused by Averaging

In the vast majority of ERP experiments, the ERP waveforms are isolated from the EEG by means of signal-averaging procedures. It is tempting to think of signal-averaging as a process that simply attenuates the nonspecific EEG, allowing us to see what the single-trial ERP waveforms look like. However, to the extent that the single-trial waveform varies from trial to trial, the averaged ERP may provide a distorted view of the single-trial waveforms, particularly when component latencies vary from trial to trial. Panels G and H of figure 2.1 illustrate this distortion. Panel G illustrates three single-trial ERP waveforms (without any EEG noise), with significant latency differences across trials, and panel H shows the average of those three single-trial waveforms. The averaged waveform differs from the single-trial waveforms in two significant ways. First, it is smaller in peak amplitude. Second, it is more spread out in time. In addition, even though the waveform in panel H is the average of the waveforms in panel G, the onset time of the averaged waveform in panel H reflects the onset time of the earliest single-trial waveform and not the average onset time. This leads to our next rule:

Rule 5. Never assume that an averaged ERP waveform accurately represents the individual waveforms that were averaged together. In particular, the onset and offset times in the averaged waveform will represent the earliest onsets and latest offsets from the individual trials or individual subjects that contribute to the average.

Fortunately, it is often possible to measure ERPs in a way that avoids the distortions created by the signal-averaging process. For example, the area under the curve in the averaged waveform shown in panel H is equal to the average of the area under the single-trial curves in panel G. In most cases, measurements of area amplitude (or mean amplitude) are superior to measurements of peak amplitude. Similarly, it is possible to find the time point that divides the area into two equal halves, and this can be a better measure of latency than peak measures (for more details, see chapter 6).

It is worth mentioning that a very large number of published ERP experiments have violated the five rules presented so far. There is no point in cataloging the cases (especially given that the list would include some of my own papers). However, violations of these rules significantly undermine the strength of the conclusions that one can draw from these experiments. For new students of the ERP technique, it would be worth reading a large set of ERP papers and trying to identify both violations of these rules and methods for avoiding the pitfalls that the rules address.

What Is an ERP Component?

So how can we accurately assess changes in latent components on the basis of the observed ERP waveforms? Ideally, we would like to be able to take an averaged ERP waveform and use some simple mathematical procedure to recover the actual waveforms corresponding to the components that sum together to create the recorded ERP waveform. We could then measure the amplitude and the latency of the isolated components, and changes in one component would not influence our measurement of the other components. Unfortunately, just as there are infinitely many generator configurations that could give rise to a given ERP scalp distribution, there are infinitely many possible sets of latent components that could be summed together to give rise to a given ERP waveform. In fact, this is the basis of Fourier analysis: any waveform can be decomposed into the sum of a set of sine waves. Similarly, techniques such as principal components analysis (PCA) and independent components analysis (ICA) use the correlational structure of a data set to derive a set of basis components that can be added together to create the observed waveforms (for more information, see Donchin & Heffley, 1978; Makeig et al., 1997). Localization techniques can also be used to compute component waveforms at the site of each ERP generator source. Unfortunately, none of these techniques has yet been perfected, as discussed in this section.

All techniques for estimating the latent components are based on assumptions about what a component is. In the early days of ERP research, a component was defined primarily on the basis of its polarity, latency, and general scalp distribution. For example, the P3a and P3b components were differentiated on the basis of the earlier peak latency and more frontal distribution of the P3a component relative to the P3b component. However, polarity, latency, and scalp distribution are superficial features that don't really capture the essence of a component. For example, the peak latency of the P3b component may vary by hundreds of milliseconds depending on the difficulty of the target-nontarget discrimination (Johnson, 1986), and the scalp distribution of the auditory N1 wave depends on the pitch of the eliciting stimulus in a manner that corresponds with the tonotopic map of auditory cortex (Bertrand, Perrin, & Pernier, 1991). Even polarity may vary: the C1 wave, which is generated in primary visual cortex, is negative for upper-field stimuli and positive for lower-field stimuli due to the folding pattern of this cortical area (Clark, Fan, & Hillyard, 1995). Consequently, many investigators now define components in terms of a combination of computational function and neuroanatomical generator site (see, e.g., Näätänen & Picton, 1987). Consistent with this approach, my own definition of the term *ERP component* is:

Scalp-recorded neural activity that is generated in a given neuroanatomical module when a specific computational operation is performed.

By this definition, a component may occur at different times under different conditions, as long as it arises from the same module and represents the same cognitive function (e.g., the encoding of an item into working memory in a given brain area may occur at different delays following the onset of a stimulus because of differences in the amount of time required to identify the stimulus and decide that it is worth storing in working memory). The scalp distribution and polarity of a component may also vary according to this definition, because the same cognitive function may occur in different parts of a cortical module under different conditions (e.g., when a visual stimulus occurs at different locations and therefore stimulates different portions of a topographically mapped area of visual cortex). It is logically possible for two different cortical areas to accomplish exactly the same cognitive process, but this probably occurs only rarely and would lead to a very different pattern of voltages, and so this would not usually be considered a single ERP component.[2]

Techniques such as PCA and ICA use the correlational structure of an ERP data set to define a set of components, and these techniques therefore derive components that are based on functional relationships. Specifically, different time points are grouped together as part of a single component to the extent they tend to vary in a correlated manner, as would be expected for time points that reflect a common cognitive process. The PCA technique, in particular, is problematic because it does not yield a single, unique set of underlying components without additional assumptions (see, e.g., Rosler & Manzey, 1981). That is, PCA really just provides a means of determining the possible set of latent component waveshapes, but additional assumptions are necessary to decide on one set of component waveshapes (and there is typically no way to verify that the assumptions are correct). The ICA technique appears to be a much better approach, because it uses both linear and nonlinear relationships to define the components. However, because ICA is a new technique, it remains to be seen whether it will turn out to be a generally useful means of identifying latent components. In particular, any technique based on identifying correlated versus independent time points will be limited by that fact that when two separate cognitive processes covary, they may be captured as part of a single component even if they occur in very different brain areas and represent different cognitive functions. For example, if all of the target stimuli in a given experimental paradigm are transferred into working memory, an ERP component associated with target detection may always be accompanied by a component associated with working memory encoding, and this may lead ICA to group them together as a single component. Moreover, both PCA and ICA fail when latencies vary across conditions. Thus, correlation-sensitive techniques may sometimes be useful for identifying latent ERP components, but it remains to be seen whether these techniques are generally effective under typical experimental conditions.

Techniques for localizing ERPs can potentially provide measures of the time course of activity within anatomically defined regions. In fact, this aspect of ERP localization techniques might turn out to be just as important as the ability to determine the neuroanatomical locus of an ERP effect. However, as discussed in chapters 1 and 7, there are no general-purpose and foolproof techniques for definitively localizing ERPs at present, and we may never have techniques that allow direct and accurate ERP localization. Thus, this approach to identifying latent ERP components is not generally practical at the present time.

In addition to mathematical techniques for determining the latent components in an ERP waveform, it is also possible to use the world's most powerful pattern analyzer, the human brain. As you gain expertise with specific components and with ERPs in general, you will be able to look at a set of ERP data and make a reasonably good inference about the underlying component structure. This involves considering how the peaks interact with each other across time and across electrode locations and coming up with an interpretation that is consistent with the observed waveforms. Kramer (1985) studied this formally, creating a set of simulated ERP waveforms and asking a set of graduate students and research assistants to make judgments about them. He found that these individuals did a good job of recovering the underlying component structure from the waveforms, but that this depended on the observer's level of experience. Thus, you should become skilled at determining the component structure of ERP waveforms over time. This is an important skill, because you will use your assessment of the latent components to guide your measurement and analysis of the waveforms.

Avoiding Ambiguities in Interpreting ERP Components

The preceding sections of this chapter are rather depressing, because it seems like there is no perfectly general means for measuring latent components from observed ERP waveforms. This is a major problem, because many ERP experiments make predictions about the effects of some experimental manipulation on a given component, and the conclusions of these experiments are valid only if the observed effects really reflect changes in that component. For example, the N400 component is widely regarded as a sensitive index of the degree of mismatch between a word and a previously established semantic context, and it would be nice to use this component to determine which of two sets of words subjects perceive as being more incongruous. If two sets of words elicit different ERP waveforms, it is necessary to know whether this effect reflects a larger N400 for one set or a larger P3 for the other set; otherwise, it is impossible to determine whether the two sets of words differ in terms of semantic mismatch or some other variable (i.e., a variable to which the P3 wave is sensitive). Here I will describe six strategies for minimizing factors that lead to ambiguous relationships between the observed ERP waveforms and the latent components.

Strategy 1. Focus on a Specific Component

The first strategy is to focus a given experiment on only one or perhaps two ERP components, trying to keep as many other components as possible from varying across conditions. If fifteen different components vary, you will have a mess, but variations in a single component are usually tractable. Of course, sometimes a "fishing expedition" is necessary when using a new paradigm, but don't count on obtaining easily interpretable results in such cases.

Strategy 2. Use Well-Studied Experimental Manipulations

It is usually helpful to examine a well-characterized ERP component under conditions that are as similar as possible to conditions in which that component has previously been studied. For example, when Marta Kutas first started recording ERPs in language paradigms, she focused on the P3 wave and varied factors such as "surprise value" that had previously been shown to influence the P3 wave in predictable ways. Of course, when she used semantic mismatch to elicit surprise, she didn't observe the expected P3 wave but instead discovered the N400 component. However, the fact that her experiments were so closely related to previous P3 experiments made it easy to determine that the effect she observed was a new negative-going component and not a reduction in the amplitude of the P3 wave.

Strategy 3.   Focus on Large Components

When possible, it is helpful to study large components such as P3 and N400. When the component of interest is very large compared to the other components, it will dominate the observed ERP waveform, and measurements of the corresponding peak in the ERP waveform will be relatively insensitive to distortions from the other components.

Strategy 4.   Isolate Components with Difference Waves

It is often possible to isolate the component of interest by creating difference waves. For example, imagine that you are interested in assessing the N400 for two different noun types, count nouns (e.g., *cup*) and mass nouns (e.g., *water*). The simple approach to this might be to present one word per second, with count nouns and mass nouns randomly intermixed. This would yield two ERP waveforms, one for count nouns and one for mass nouns, but it would be difficult to know if any differences observed between the count noun and mass noun waveforms were due to difference in N400 amplitude or due to differences in some other ERP component.

To isolate the N400, you could redesign the experiment so that each trial contained a sequence of two words, a context word and a target word, with count noun target word on some trials and a mass noun target word on others. In addition, the context and target words would sometimes be semantically related and sometimes be semantically unrelated. You would then have four types of trial types:

Count noun, related to context word (e.g., "plate … cup")
Mass noun, related to context word (e.g., "rain … water")
Count noun, unrelated to context word (e.g., "sock … cup")
Mass noun, unrelated to context word (e.g., "garbage … water")

You could then isolate the N400 by constructing difference waves in which the ERP waveform elicited by a given word when it was preceded by a semantically related context word is subtracted from the ERP waveform elicited by that same word when preceded by a semantically unrelated context word. Separate difference waves would be constructed for count nouns and mass nouns (unrelated minus related count nouns and unrelated minus related mass nouns). Each of these difference waves should be dominated by a large N400 component, with little or no contribution from other components (because most other components aren't sensitive to semantic mismatch). You could then see if the N400 was larger in the count noun difference wave or in the mass noun difference wave (I describe a real application of this general approach at the very end of this chapter—see Vogel, Luck, & Shapiro, 1998 for details).

Although this approach is quite powerful, it has some limitations. First, difference waves constructed in this manner may contain more than one ERP component. For example, there may be more than one ERP component that is sensitive to the degree of semantic mismatch, so an unrelated-minus-related difference wave might consist of two or three components rather than just one. However, this is still a vast improvement over the raw ERP waveforms, which probably contain at least ten different components. The second limitation of this approach is that it is sensitive to interactions between the variable of interest (e.g., count nouns versus mass nouns) and the factor that is varied to create the difference waves (e.g., semantically related versus unrelated word pairs). If, for example, the N400 amplitude is 1 µV larger for count nouns than for mass nouns, regardless of the degree of semantic mismatch, then count noun difference waves will be identical to the mass noun difference waves. Fortunately, when two factors influence the same ERP component, they are likely to interact multiplicatively. For example, N400 amplitude might be 20 percent greater for count nouns than for mass nouns, leading to a larger absolute difference in N400 amplitude when the words are unrelated to the context word than when they are related. Of course, the interactions could take a more complex form that would lead to unexpected results. For example, count nouns could elicit a larger N400 than mass nouns when the words are unrelated to the context word, but they might elicit a smaller N400 when the words are related to the context word. Thus, although difference waves can be very helpful in isolating specific ERP components, care is necessary when interpreting the results.

I should also mention that the signal-to-noise ratio of a difference wave will be lower than those of the original ERP waveforms. Specifically, if the original waveforms have similar noise levels, then the noise in the difference wave will be larger by a factor of the square root of two (i.e., approximately 40 percent larger).

Strategy 5.   Focus on Components That Are Easily Isolated

The previous strategy advocated using difference waves to isolate ERP components, and one can refine this by focusing on certain ERP components that are relatively easy to isolate. The best example of this is the lateralized readiness potential (LRP), which reflects movement preparation and is distinguished by its contralateral scalp distribution. Specifically, the LRP in a given hemisphere is more negative when a movement of the contralateral hand is being prepared than when a movement of the ipsilateral hand is being prepared, even if the movements are not executed. In an appropriately designed experiment, only the motor preparation will lead to lateralized ERP components, making it possible to form difference waves in which all ERPs are subtracted away except for those related to lateralized motor preparation (see Coles, 1989; Coles et al., 1995). Similarly, the N2pc component for a given hemisphere is more negative when attention is directed to the contralateral visual field than when it is directed to the ipsilateral field, even when the evoking stimulus is bilateral. Because most of the sensory and cognitive components are not lateralized in this manner, the N2pc can be readily isolated (see, e.g., Luck et al., 1997; Woodman & Luck, 2003).

Strategy 6.   Component-Independent Experimental Designs

The best strategy is to design experiments in such a manner that it does not matter which latent ERP component is responsible for the observed changes in the ERP waveforms. For example, Thorpe and colleagues (1996) conducted an experiment in which they asked how quickly the visual system can differentiate between different classes of objects. To answer this question, they presented subjects with two classes of photographs, pictures that contained animals and pictures that did not. They found that the ERPs elicited by these two classes of pictures were identical until approximately 150 ms, at which point the waveforms diverged. From this experiment, it is possible to infer that the brain can detect the presence of an animal in a picture by 150 ms, at least for a subset of pictures (note that the onset latency represents the trials and subjects with the earliest onsets and not necessarily the average onset time). This experimental effect occurred in the time range of the N1 component, but it may or may not have been a modulation of that component. Importantly, the conclusions of this study do not depend at all on which latent component the experimental manipulation influenced. Unfortunately, it is rather unusual to be able to answer a significant question in cognitive neuroscience using ERPs in a component-independent manner, but one should use this approach whenever possible. I will provide additional examples later in this chapter.

Avoiding Confounds and Misinterpretations

The problem of assessing latent components on the basis of observed ERP waveforms is usually the most difficult aspect of the design and interpretation of ERP experiments, and this problem is particularly significant in ERP experiments. There are other significant experimental design issues that are applicable to a wide spectrum of techniques, but are particularly salient in ERP experiments; these will be the focus of this section.

The most fundamental principle of experimentation is to make sure that a given experimental effect has only a single possible cause. One part of this principle is to avoid confounds, but a subtler part is to make sure that the experimental manipulation doesn't have secondary effects that are ultimately responsible for the effect of interest. For example, imagine that you observe that the mass of a beaker of hot water is less than the mass of a beaker of cold water. This might lead to the erroneous conclusion that hot water has a lower mass than cool water, even though the actual explanation is that some of the heated water turned to steam, which escaped through the top of the beaker. To reach the correct conclusion, it is necessary to seal the beakers so that water does not escape. Similarly, it is important to ensure that experimental manipulations in ERP experiments do not have unintended side effects that lead to an incorrect conclusion.

To explore the most common problems that occur in ERP experiments, let's consider a thought experiment that examines the effects of stimulus discriminability on P3 amplitude. In this experiment, letters of the alphabet are presented foveally at a rate of one per second and the subject is required to press a button whenever the letter Q is presented. A Q is presented on 10 percent of trials and a randomly selected non-Q letter is presented on the other 90 percent. In addition, the letter Q never occurs twice in succession. In one set of trial blocks, the stimuli are bright and therefore easy to discriminate (the bright condition), and in another set of trial blocks the stimuli are very dim and therefore difficult to discriminate (the dim condition).

There are several potential problems with this seemingly straightforward experimental design, mainly due to the fact that the target letter (Q) differs from the nontarget letters in several ways. First, the target category occurs on 10 percent of trials whereas the nontarget category occurs on 90 percent of trials. This is one of the two intended experimental manipulations (the other being target discriminability). Second, the target and nontarget letters are different from each other. Not only is the target letter a different shape from the nontarget letters—and might therefore elicit a somewhat different ERP waveform—the target letter also occurs more frequently than any of the individual nontarget letters. To the extent that the visual system exhibits long-lasting and shape-specific adaptation to repeated stimuli, it is possible that the response to the letter Q will become smaller than the response to the other letters. These physical stimulus differences probably won't have a significant effect on the P3 component, but they could potentially have a substantial effect on earlier components (for a detailed example, see experiment 4 of Luck & Hillyard, 1994a).

An important principle of ERP experimental design that I learned from Steve Hillyard is that you should be very careful to avoid confounding differences in psychological factors with subtle differences in stimulus factors. In this example experiment, for example, the probability difference between the target and nontarget letters—which is intended to be a psychological manipulation— is confounded with shape differences and differences in sensory adaptation. Although such confounds primarily influence sensory responses, these sensory responses can last for hundreds of milliseconds, and you will always have the nagging suspicion that your P3 or N400 effects reflect a sensory confound rather than your intended psychological manipulation. The solution to this confound is to make sure that your manipulations of psychological variables are not accompanied by any changes in the stimuli (including the sequential context of the stimulus of interest). This is typically accomplished by using the same stimulus sequences in two or more conditions and using verbal or written instructions to manipulate the subject's task. A statement that is prominently displayed on a wall in my laboratory summarizes this:

The Hillyard Principle—Always compare ERPs elicited by the same physical stimuli, varying only the psychological conditions.

Of course, implementing the Hillyard Principle is not always possible. For example, you may want to examine the ERPs elicited by closed-class words (articles, prepositions, etc.) and open-class words (nouns, verbs, etc.); these are by definition different stimuli. But you should be very careful that any ERP differences reflect the psychological differences between the stimuli and not low-level physical differences (e.g., word length). I can tell you from experience that every time I have violated the Hillyard Principle, I have later regretted it and ended up running a new experiment. Box 2.1 provides two examples of confounds that somehow crept into my own experiments.

A third difference between the target and nontarget letters is that subjects make a response to the targets and not to the nontargets. Consequently, any ERP differences between the targets and nontargets could be contaminated by motor-related ERP activity. A fourth difference between the targets and the nontargets is that because the target letter never occurred twice in succession, the target letter was always preceded by a nontarget letter, whereas nontarget letters could be preceded by either targets or nontargets. This is a common practice, because the P3 to the second of two targets tends to be reduced in amplitude. This is usually a bad idea, however, because the response to a target is commonly very long-lasting and extends past the next stimulus and therefore influences the waveform recorded for the next stimulus. Thus, there may appear to be differences between the target and nontarget waveforms in the N1 or P2 latency ranges that actually reflect the offset of the P3 from the previous trial, which is present only in the nontarget waveforms under these conditions. This type of differential overlap occurs in many ERP experiments, and it can be rather subtle. (For an extensive discussion of this issue, see Woldorff, 1988.)

A fifth difference between the targets and the nontargets arises when the data are averaged and a peak amplitude measure is used to assess the size of the P3 wave. Specifically, because there are many more nontarget trials than target trials, the signal-to-noise ratio is much better for the nontarget waveforms. The maximum amplitude of a noisy waveform will tend to be greater than the

Box 2.1   Examples of Subtle Confounds

It may seem simple to avoid physical stimulus confounds, but these confounds are often subtle. Let me give two examples where I made mistakes of this nature. Many years ago, I conducted a series of experiments in which I examined the ERPs elicited by visual search arrays consisting of seven randomly positioned "distractor" bars of one orientation and one randomly positioned "pop-out" bar of a different orientation. In several experiments, I noticed that the P1 wave tended to be slightly larger over the hemisphere contralateral to the pop-out item relative to the ipsilateral hemisphere. I thought this might reflect an automatic capture of attention by the pop-out item, although this didn't fit very well with what we knew about the time course of attention. Fortunately, Marty Woldorff suggested that this effect might actually reflect a physical stimulus confound. Specifically, the location of the pop-out bar on one trial typically contained an opposite-orientation distractor bar on the previous trial, whereas the location of a distractor bar on one trial typically contained a same-orientation distractor bar on the previous trial. Thus, the response to the pop-out bar might have been larger than the response to the distractor bars because the neurons coding the pop-out bar wouldn't have just responded to a stimulus of the same orientation, and this led to a larger response contralateral to the pop-out bar because these responses are usually larger at contralateral than at ipsilateral sites. At first, this seemed unlikely because the screen was blank for an average of 750 ms between trials. However, when I tested Marty's hypothesis with an additional experiment, it turned out that he was correct (see experiment 4 of Luck & Hillyard, 1994a).

I encountered another subtle physical stimulus confound several years later when Massimo Girelli and I were examining whether the N2pc component in a visual search task differed for pop-out stimuli in the upper visual field compared to the lower visual field. Because this question directly involves a physical stimulus manipulation, we knew we had to use a trick to avoid direct physical stimulus confounds. The trick was to use two differently colored pop-out items on each trial, one in the left visual field and one in the right visual field, and to ask the subjects to attend to one color (e.g., the red pop-out) on some trial blocks and to attend to the other color (e.g., the green pop-out) on others. Because the N2pc component is lateralized with respect to the direction of attention, we could then make a difference wave in which the physical stimulus was held constant (e.g., red pop-out on the left and green pop-out on the right) and attention was varied (e.g., attend-red versus attend-green). We expected that all of the purely sensory responses would be subtracted away in this difference wave because the physical stimulus was identical and only attention was varied. This is a useful trick that we have

Box 2.1 (continued)

used several times, but in this case we failed to implement it correctly. Specifically, the goal of our experiment was to compare the responses to upper versus lower field stimuli, and we therefore subdivided our trials as a function of the vertical position of the target item. Unfortunately, when the target was in the upper field, the nontarget pop-out item could be either in the upper field or the lower field. We had coded the vertical position of the attended pop-out item but not the vertical position of the unattended pop-out item, and it was therefore impossible to perform subtractions on exactly the same physical stimuli (e.g., red in upper-left and green in upper-right when red was attended versus when green was attended). When we performed subtractions that were not properly controlled, we obtained a pattern of results that was simply weird, and we had to run the experiment a second time with the appropriate codes.

maximum amplitude of a clean waveform due purely to probability, and a larger peak amplitude for the target waveform could therefore be caused solely by its poorer signal-to-noise ratio even if the targets and nontargets elicited equally large responses.

The manipulation of stimulus brightness is also problematic, because this will influence several factors in addition to stimulus discriminability. First, the brighter stimuli are, well, brighter than the dim stimuli, and this may create differences in the early components that are not directly related to stimulus discriminability. Second, the task will be more difficult with the dim stimuli than with the bright stimuli. This may induce a greater state of arousal during the dim blocks than during the bright blocks, and it may also induce strategy differences that lead to a completely different set of ERP components in the two conditions. A third and related problem is that reaction times will be longer in the dim condition than in the bright condition, and any differences in the ERP waveforms between these two conditions could be due to differences in the time course of motor-related ERP activity (which overlaps with the P3 wave).

There are two main ways to overcome problems such as these. First, you can avoid many of these problems by designing the experiment differently. Second, it is often possible to demonstrate that a potential confound is not actually responsible for the experimental effect; this may involve additional analyses of the data or additional experiments. As an illustration, let us consider several steps that you could take to address the potential problems in P3 experiment described above:

1. You could use a different letter as the target for each trial block, so that across the entire set of subjects, all letters are approximately equally likely to occur as targets or nontargets. This solves the problem of having different target and nontarget shapes.

2. To avoid differential visual adaptation to the target and nontarget letters, you could use a set of ten equiprobable letters, with one serving as the target and the other nine serving as nontargets. Each letter would therefore appear on 10 percent of trials. If it is absolutely necessary that one physical stimulus occurs more frequently than another, it is possible to conduct a sequential analysis of the data to demonstrate that differential adaptation was not present. Specifically, trials on which a nontarget was preceded by a target can be compared with trials on which a nontarget was preceded by a nontarget. If no difference is obtained—or if any observed differences are unlike the main experimental effect—then the effects of stimulus probability are probably negligible.

3. Rather than asking the subjects to respond only to the targets, instruct them to make one response for targets and another for nontargets. Target and nontarget RTs are likely to be different, so some differential motor activity may still be present for targets versus nontargets, but this is still far better than having subjects respond to the targets and not to the nontargets.

4. It would be a simple matter to eliminate the restriction that two targets cannot occur in immediate succession, thus avoiding the possibility of differential overlap from the preceding trial. However, if it is necessary to avoid repeating the targets, it is possible to con-

struct an average of the nontargets that excludes trials preceded by a target. If this is done, then both the target and the nontarget waveforms will contain only trials on which the preceding trial was a nontarget (as in step 2).

5. There are two good ways to avoid the problem that peak amplitudes tend to be larger when the signal-to-noise ratio is lower. First, as discussed above, the peak of an ERP waveform bears no special relationship to the corresponding latent component, so this problem can be solved by measuring the mean amplitude over a predefined latency range rather than the peak amplitude. As discussed in chapter 4, mean amplitude has several advantages over peak amplitude, and one of them is that the measured amplitude is not biased by the number of trials. Of course, mean amplitude measures exhibit increased variance as the signal-to-noise ratio decreases, but the expected value does not vary. If, for some reason, it is necessary to measure peak amplitude rather than mean amplitude, it is possible to avoid biased amplitude measures by creating the nontarget average from a randomly selected subset of the nontarget trials such that the target and nontarget waveforms reflect the same number of trials.

6. There is no simple way to compare the P3 elicited by bright stimuli versus dim stimuli without contributions from simple sensory differences. However, simple contributions can be ruled out by a control experiment in which the same stimuli are used but are viewed during a task that is unlikely to elicit a P3 wave (e.g., counting the total number of stimuli, regardless of the target-nontarget category). If the ERP waveforms for the bright and dim stimuli in this condition differ only in the 50–250 ms latency range, then the P3 differences observed from 300–600 ms in the main experiment cannot easily be explained by simple sensory effects and must instead reflect an interaction between sensory factors (e.g., perceptibility) and cognitive factors (e.g., whatever is responsible for determining P3 amplitude).

7. The experiment should also be changed so that the bright and dim stimuli are randomly intermixed within trial blocks. In this way,

the subject's state at stimulus onset will be exactly the same for the easy and difficult stimuli. This also tends to reduce the use of different strategies.

8. It is possible to use additional data analyses to test whether the different waveforms observed for the dim and bright conditions are due to differences in the timing of the concomitant motor potentials (which is plausible whenever RTs differ between two conditions). Specifically, if the trials are subdivided into those with fast RTs and those with slow RTs, it is possible to assess the size and scalp distribution of the motor potentials. If the difference between trials with fast and slow RTs is small compared to the main experimental effect, or if the scalp distribution of the difference is different from the scalp distribution of the main experimental effect, then this effect probably cannot be explained by differential motor potentials.

Most of these strategies are applicable in many experimental contexts, and they reflect a set of general principles that are very widely applicable. I will summarize these general principles in some additional rules:

Rule 6. Whenever possible, avoid physical stimulus confounds by using the same physical stimuli across different psychological conditions (the Hillyard Principle). This includes "context" confounds, such as differences in sequential order.

Rule 7. When physical stimulus confounds cannot be avoided, conduct control experiments to assess their plausibility. Never assume that a small physical stimulus difference cannot explain an ERP effect.

Rule 8. Be cautious when comparing averaged ERPs that are based on different numbers of trials.

Rule 9. Be cautious when the presence or timing of motor responses differs between conditions.

Rule 10. Whenever possible, experimental conditions should be varied within trial blocks rather than between trial blocks.

Examples from the Literature

In this section, I will discuss three experiments that used ERPs to answer a significant question in cognitive neuroscience. In each case, the experiments were designed in such a manner that the conclusions did not depend on identifying specific ERP components, making the results much stronger than is otherwise possible.

Example 1:   Auditory Selective Attention

Background   Given that I did my graduate work with Steve Hillyard, I was imprinted at an early age on the experimental design that he and his colleagues developed in 1973 to address the classic "locus-of-selection" question. This question asks whether attention operates at an early stage of processing, allowing only selected inputs to be perceived and recognized (the *early selection* position of investigators such as Broadbent, 1958; Treisman, 1969), or whether attention instead operates at a late stage such that all incoming sensory events receive equal perceptual processing but only selected stimuli reach decision processes, memory, and behavioral output systems (the *late selection* position of investigators such as Deutsch & Deutsch, 1963; Norman, 1968).

Two factors have made this a difficult question to address with traditional behavioral measures. First, it is difficult to assess the processing of an ignored stimulus without asking the subjects to respond to it, in which case the stimulus may become attended rather than ignored. Second, if responses to ignored stimuli are slower or less accurate than responses to attended stimuli, it is difficult to determine whether this reflects an impairment in sensory processes or an impairment of higher level decision, memory, or response processes. ERPs, in contrast, are particularly well suited for solving both of these problems. First, ERPs are easily recorded in the absence of an overt response, making them ideal for monitoring the processing of an ignored stimulus. Second, ERPs provide precise information about the time course of processing, making them ideal for answering questions about the stage of processing at which an experimental effect occurs. In the case of attention, for example, it is possible to determine whether the early sensory-evoked ERP components are suppressed for ignored stimuli relative to attended stimuli, which would be consistent with the early selection hypothesis, or whether the early ERP components are identical for attended and ignored stimuli and only the later ERP components are sensitive to attention.

Experimental Design   Several studies in the 1960s and early 1970s used this logic to address the locus-of-selection question, but these experiments had various methodological shortcomings that made them difficult to interpret. Hillyard, Hink, Schwent, and Picton (1973) reported an experiment that solved these problems and provided unambiguous evidence for early selection. Figure 2.2A illustrates this experiment. Subjects were instructed to attend to the left ear in some trial blocks and the right ear in others. A rapid sequence of tone pips was then presented, with half of the stimuli presented in each ear. To make the discrimination between the two ears very easy, the tones were presented at a different pitch in each ear. Subjects were instructed to monitor the attended ear and press a button whenever a slightly higher pitched "deviant" target tone was detected in that ear, which occurred infrequently and unpredictably. Higher pitched tones were also presented occasionally in the ignored ear, but subjects were instructed not to respond to these "ignored deviants."

In some prior ERP experiments, subjects were required to respond to or count all attended stimuli and withhold responses to all ignored stimuli (including internal responses such as counting). As a result, any differences in the ERPs evoked by attended and ignored stimuli could have been due to response-related activity that was present for the attended ERPs but absent for ignored ERPs (see rule 9). Hillyard et al. (1973) avoided this problem by presenting both target and nontarget stimuli in both the attended and ignored ears and asking the subjects to count the targets in

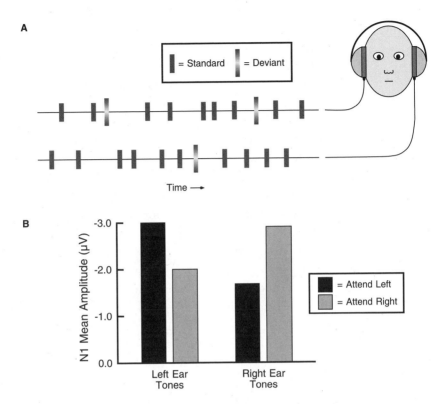

Figure 2.2 Experimental paradigm (A) and results (B) from the study of Hillyard et al. (1973). Subjects listened to streams of tone pips in the left ear and right ear. Most of the tones were a standard frequency (800 Hz in the left ear and 1500 Hz in the right ear), but occasional deviant tones were presented at a slightly higher frequency (840 Hz left; 1560 Hz right). Subjects were instructed to attend to the left ear for some trial blocks and to the right ear for others, and counted the number of deviant tones in the attended ear. The average N1 amplitude was measured for the standard tones and used as an index of sensory processing. Left ear tones elicited a larger N1 wave when attention was directed to the left ear than when attention was directed to the right ear; conversely, right ear tones elicited a larger N1 wave when attention was directed to the right ear than when attention was directed to the left ear. (Reproduced by permission from Luck, 1998a. © 1998 Psychology Press.)

the attended ear. The analyses were focused on the nontargets, to which subjects made neither an external response such as a button press nor an internal response such as counting. To ensure that subjects attended to the nontargets in the attended ear, even though no response was required for them, targets and nontargets within an ear were difficult to discriminate from each other (but easy to discriminate from the stimuli in the other ear). Thus, subjects were motivated to focus attention onto all of the stimuli presented in one ear and ignore all stimuli within the other ear. The main experimental question was whether the early sensory ERP components evoked by a nontarget stimulus presented in the attended ear would be larger than those evoked by a nontarget stimulus presented in the ignored ear.

The sensory ERP components are highly sensitive to the physical characteristics of the evoking stimulus. As a result, one cannot legitimately compare the ERP evoked by an attended tone in the left ear with an ignored tone in the right ear: any differences between these ERPs could be due to differences between the two ears that have nothing to do with attention (see rule 6). The design Hillyard et al. (1973) employed circumvents this problem by allowing the ERP elicited by the same physical stimulus to be compared under different psychological conditions. For example, one can compare the ERP evoked by a left nontarget during attend-left blocks with the ERP evoked by the same left nontarget during attend-right blocks. Because both cases use the same stimulus, any differences in the ERPs between the attend-left and attend-right conditions must be due to differences in attentional processing.

In some attention experiments, the investigators compare an "active" condition, in which the subject responds to the stimuli, with a "passive" condition, in which the subject makes no response to the stimuli and perhaps engages in a distracting activity such as reading a book. Frequently, however, the task in the active condition is much more demanding than the distraction task in the passive condition, leading to greater overall arousal during the active condition. If we compare the ERPs in the two conditions, any differences might be due to these global arousal differences rather than selective changes in stimulus processing. This is a violation of rule 10. To ensure that differences in global arousal would not interfere with their study, Hillyard et al. (1973) compared ERPs evoked during equally difficult attend-left and attend-right conditions rather than active and passive conditions.

Results   Now that we have discussed the logic behind this study, let us consider the results. As figure 2.2B shows, the N1 component was found to be larger for attended stimuli than for ignored stimuli.[3] Specifically, the N1 elicited by left ear tones was larger when the left ear was attended than when the right ear was attended, and the N1 elicited by right ear tones was larger when the right ear was attended than when the left ear was attended. These effects began approximately 60–70 ms after stimulus onset and peaked at approximately 100 ms poststimulus. In this manner, Hillyard et al. (1973) were able to demonstrate that, at least under certain conditions, attention can influence the processing of a stimulus within the first 100 milliseconds after stimulus onset, which is consistent with the early selection hypothesis.

Larger Issues   I would like to emphasize three aspects of this study. First, it was specifically designed to address an existing and significant question that had previously eluded investigators. This contrasts with many ERP experiments in which it seems as if the authors simply took an interesting cognitive paradigm and ran it while recording ERPs to "see what happens." This approach, although sometimes a useful first step, rarely leads to important conclusions about cognitive or neural issues. Second, this study does not rely on identifying specific ERP components (see strategy 6). The ERPs elicited by the attended and ignored stimuli diverged around 60 to 70 ms poststimulus, and it is this timing information that is the crucial result of the study, and not the fact that the effect occurred in the latency range of the N1 component. In fact, there has been much controversy about whether attention influences the

N1 component per se, but the finding of an effect within 100 ms of stimulus onset will continue to be important regardless of the outcome of this dispute.

A third important aspect of this study is that it used ERPs to assess the processing of stimuli for which subjects made no overt response. This is one of the main advantages of the ERP technique over behavioral measures, and many of the most significant ERP experiments have exploited this ability of ERPs to be used for the "covert monitoring" of cognitive processes.

Example 2:   Partial Information Transmission

Background   Most early models of cognition assumed that simple cognitive tasks were accomplished by means of a sequence of relatively independent processing stages. These models were challenged in the late 1970s by models in which different processes could occur in parallel, such as McClelland's (1979) cascade model and Eriksen and Schultz's (1979) continuous flow model. At the most fundamental level, these new models differed from the traditional models in that they postulated that partial results from one process could be transmitted to another process, such that the second process could begin working before the first process was complete. Traditional discrete-stage models, in contrast, assumed that a process worked until it achieved its result, which was then passed on to the next stage. This is a crucial distinction, but it is very difficult to test without a direct means of observing the processes that are thought to overlap. This is exactly the sort of issue that ERPs can easily address.

By coincidence, two different laboratories conducted similar ERP studies of this issue at about the same time, and both used the lateralized readiness potential (LRP) as a means of assessing whether response systems can become activated before stimulus identification is complete (Miller & Hackley, 1992; Osman et al., 1992). Both were excellent studies, but here I will focus on the one conducted by Miller and Hackley. These investigators tested the specific hypothesis that subjects will begin to prepare a response to a stimulus based on the most salient aspects of the stimulus, even if they later withheld that response because of additional information extracted from the stimulus. In other words, they predicted that motor processing may sometimes begin before perceptual processing is complete, which would be incompatible with traditional discrete-stage models of cognition.

Experimental Design   To test this hypothesis, they presented subjects with one of four stimuli on each trial: a large S; a small S; a large T; or a small T. Subjects responded with one hand for S and with the other hand for T, but they responded only to one of the two sizes; they gave no response for the other size (half of the subjects responded to large stimuli and half to small). Thus, this paradigm was a hybrid of a go/no-go design (go for one size and no-go for the other) and a two-alternative forced choice design (one response for S, a different response for T). The shape difference was very salient, but the size difference was relatively difficult to discriminate. Consequently, subjects could begin to prepare a given hand to respond as soon as they discriminated the shape of the letter, and they could later choose to emit or withhold this response when they eventually discriminated the size of the letter.

To determine if the subjects prepared a specific hand for response on the basis of letter shape, even on trials when they made no response because the letter was the wrong size, Miller and Hackley used the LRP component. The LRP component, which reflects response preparation, is particularly useful for this purpose because it is lateralized with respect to the responding hand. Miller and Hackley's (1992) paper provides a wonderful review of the LRP literature, and I highly recommend reading it. Here I will mention two essential aspects of the LRP that were essential for testing their hypothesis. First, the LRP begins before muscle contractions begin and can occur in the absence of an overt response, indicating that it reflects response preparation. Second, the LRP is larger over the hemisphere contralateral to the movement being prepared. Consequently, the presence of an LRP is virtually absolute proof that the brain has begun to differentially prepare one hand for responding. In other words, there is no way to get a consistently larger response over the hemisphere contralateral to a given hand unless the brain has begun to prepare a response for that hand. Thus, Miller and Hackley predicted that an LRP would be observed contralateral to the hand indicated by the shape of the letter, even if the size of the letter indicated that no response should be made (note that EMG recordings were used to ensure that absolutely no response was made on these trials). When the size of the letter was finally perceived, the LRP would then be terminated if it indicated that no response should be made.

Before I describe the results, I want to mention a technique to isolate the LRP component and to ensure that it purely reflects contralateral hand preparation. Imagine that we compare the ERPs recorded at left and right premotor electrode sites just prior to a left-hand response. If the voltage were more negative over the right hemisphere than the left hemisphere, we might conclude that the brain was preparing the left-hand response, leading to a greater negativity over the right hemisphere. This conclusion would be unwarranted, however, because the same result would be obtained if the right hemisphere yielded a more negative response regardless of which hand was being prepared. Similarly, imagine that we compare the response to left-hand and right-hand responses at a left premotor electrode site. If the voltage were more negative prior to left-hand responses than prior to right-hand responses, we might again conclude the brain was preparing the left-hand response. This would again be unwarranted, because the same result would be obtained if left-hand responses yielded more negative voltages than right-hand responses over both hemispheres.

To isolate activity that purely reflects lateralized response preparation, it is necessary to record waveforms corresponding to the different combinations of left and right hemisphere electrode (E_{left} and E_{right}) and left and right hand response (R_{left} and R_{right}). There are four combinations: left hemisphere electrode with left hand response ($E_{left}R_{left}$), left hemisphere electrode with right hand response ($E_{left}R_{right}$), right hemisphere electrode with left hand response ($E_{right}R_{left}$), and right hemisphere electrode with right hand response ($E_{right}R_{right}$). As discussed by Coles (1989), the LRP can then be isolated by creating a difference wave according to this formula:

$$LRP = [(E_{right}R_{left} - E_{left}R_{left}) + (E_{left}R_{right} - E_{right}R_{right})] \div 2 \qquad (2.1)$$

This formula computes the average of the contralateral-minus-ipsilateral difference for left-hand and right-hand responses, eliminating any overall differences between the left and right hemispheres (independent of hand) and any overall differences between the left and right hands (independent of hemisphere). All that remains is the extent to which the response is generally larger over the hemisphere contralateral to the hand being prepared, regardless of which hand is being prepared or which hemisphere is being recorded. This procedure effectively isolates the LRP from the vast majority of ERP components, which do not show this intricate interaction between hemisphere and hand (see strategy 5).

Results   Figure 2.3 shows the results of this study. Panel A shows the averaged ERP waveforms from the left- and right-hemisphere electrodes sites for left- and right-hand responses when the stimulus was the appropriate size for a response ("Go Trials"). The electrode sites used here are labeled C3′ and C4′ to indicate premotor sites just lateral to the standard C3 and C4 sites in the left and right hemispheres, respectively. The ERP waveform at the left-hemisphere site was more negative on trials with a right-hand response than on trials with a left-hand response, and the complementary pattern was observed at the right-hemisphere site. This effect began approximately 200 ms poststimulus and continued for at least 800 ms, which is typical for the LRP component.

Panel B of figure 2.3 shows the responses observed when the size of the stimulus indicated that no response should be made ("No-Go

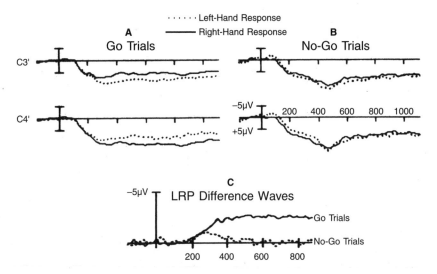

Figure 2.3   Grand-average ERP waveforms from the study of Miller and Hackley (1992). The waveforms in panels A and B reflect trials on which the size of the letter indicated that a response should be made ("Go Trials," panel A) or should not be made ("No-Go Trials," panel B). These waveforms were recorded from left central (C3') and right central (C4') electrode sites. Broken lines reflect trials on which the shape of the letter indicated a left-hand response, and solid lines reflect trials on which the shape of the letter indicated a right-hand response. Panel C shows LRP difference waves for the go and no-go trials. Note that a brief LRP deflection was present on no-go trials, even though no response (or EMG activity) was present on these trials. Negative is plotted upward. (Adapted with permission from Miller and Hackley, 1992. © 1992 American Psychological Association.)

Trials"). From approximately 200–400 ms poststimulus, the ERPs were more negative over the left hemisphere when the shape of the stimulus was consistent with the right-hand response than with the left-hand response, and the complementary pattern was observed at the right-hemisphere site. After 400 ms, however, the ERP waveform was slightly more negative in both hemispheres for right-hand responses relative to left-hand responses.

Panel C shows the waveforms for the Go and No-Go trials after the LRP was isolated by means of equation 2.1. For both Go and No-Go trials, the LRP began to deviate from baseline at approximately 200 ms. On Go trials, the LRP continued until the end of the recording epoch. On No-Go trials, in contrast, the LRP returned to baseline at approximately 400 ms. Thus, the brain began to prepare the response indicated by the shape of the letter on both Go and No-Go trials, even though no response was executed on No-Go trials. This provides what I believe to be ironclad evidence that, at least under some conditions, response systems receive partial information about a stimulus before the stimulus has been fully identified.

Larger Issues   There are many unjustified conclusions that you might be tempted to draw from the waveforms shown in figure 2.3. First, you might suppose that subjects typically began preparing the response at approximately 200 ms poststimulus, the time point at which the LRP began to deviate from baseline. However, as discussed previously in this chapter, the onset of a response in an averaged ERP waveform reflects the earliest onset times, not the average onset times (see rule 5). Thus, the 200-ms onset latency of the LRP in figure 2.3C reflects the fastest trials from the fastest subjects. Similarly, you might be tempted to assume that the LRP was only about half as large on No-Go trials as on Go trials. However, it is possible that the single-trial LRPs were just as large on No-Go trials as on Go trials, but were present on only 50 percent of trials.

Fortunately, the main conclusions from this experiment do not depend on any unjustifiable conclusions. In fact, it doesn't even matter whether or not the waveforms shown in figure 2.3C reflect the same LRP component that was observed in previous experiments. The simple fact that the hemispheric distribution of the voltage was different when the stimulus signaled a contralateral response rather than an ipsilateral response is sufficient to provide solid evidence that the brain had begun to determine which response was associated with the shape of the letter. So it doesn't matter if the effect reflects an ipsilaterally larger P3 component, a contralaterally larger N400 component, or some new, never-before-observed component (see strategy 6).

I would like to make one additional observation about this experiment: the data in figure 2.3 are extremely clean. In the raw ERP waveforms (panels A and B), the waveforms are almost completely noise-free during the prestimulus interval (the noise level in the prestimulus interval is very useful for assessing the quality of an ERP waveform). Even in the difference waves (panel C), which are plotted at a higher magnification, the prestimulus noise level is very small compared to the size of the LRP effects. Clean data lead to much greater confidence and much stronger conclusions; even if the p-value of an experimental effect passes the magical .05 criterion, noisy looking waveforms make a poor impression and make it difficult to have confidence in the details of the results (e.g., the onset time of an effect).

The following chapters will provide many hints for recording clean data, but I want to emphasize one factor here: clean ERP waveforms require a very large number of trials. For example, I usually try to have ten to twenty subjects in an experiment, and for each type of trial in the experiment I collect approximately sixty trials per subject when I'm examining a large component (e.g., P3 or N400), about 150 trials per subject for a medium-sized component (e.g., N2), and about 400 trials per subject for a small

Box 2.2   My Favorite ERP Component

Although I have never conducted an LRP experiment, I think the LRP may be the single most useful ERP component for addressing a broad spectrum of cognitive issues at this time. First, the LRP is very easy to isolate from other ERP components by means of difference waves. Second, it has been very well characterized, including its neural generator sources (for a review, see Coles, 1989). Third, it is relatively easy to design an experiment in such a way that you can isolate the LRP component (all you really need is to have some stimuli indicate left-hand responses and others indicate right-hand responses, with the assignment of stimuli to hands counterbalanced across trial blocks). Finally, the LRP has been used to provide solid answers to a variety of interesting questions (e.g., de Jong et al., 1990; Dehaene et al., 1998; Miller & Hackley, 1992; Osman & Moore, 1993) (see also review by Coles et al., 1995).

component (e.g., P1). This usually requires a session of three to four hours (including the time required to apply the electrodes and train the subject to perform the task). The need for large numbers of trials in ERP experiments is unfortunate, because it limits the questions that this technique can realistically answer. However, there isn't much point in conducting an experiment that would be very interesting except that the results are so noisy that they are difficult to interpret.

Example 3:   Dual-Task Performance

Background   It is often difficult for a person to perform two tasks at the same time. For example, it is difficult to discuss the principles of ERP experimental design while driving a BMW 325xi at high speeds on a winding road in a snowstorm. Dual-task interference can also be seen in much simpler tasks, and many researchers have recently studied dual-task interference in the *attentional blink* paradigm (see Shapiro, Arnell, & Raymond, 1997 for a review). In this paradigm, a very rapid stream of approximately twenty stimuli is presented at fixation on each trial, and the subject is required to detect two targets (called T1 and T2) from among the nontarget stimuli. Figure 2.4A illustrates a typical experiment. In this experiment, T1 and T2 are digits and the nontarget stimuli are letters. At the end of each trial, subjects report the identities of the two digits. While subjects are processing T1, they may be unable to effectively process T2, leading to errors in identifying T2. This would be expected to occur primarily when T2 is presented shortly after T1, and to assess the time course of the interference between T1 and T2, the lag between T1 and T2 is varied.

Figure 2.4B shows the pattern of results that is typically observed in this paradigm. T2 accuracy is typically severely impaired when T2 occurs two to four items after T1, but T2 accuracy is quite high at a lag of one item or at lags of five or more items. In contrast, T1 accuracy is typically quite good at all lags. This pattern of results is

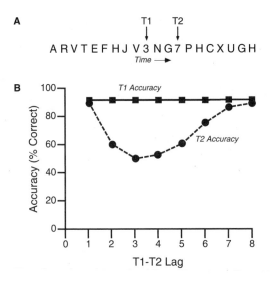

A

T1   T2

A R V T E F H J V 3 N G 7 P H C X U G H

Time ⟶

B

Figure 2.4 Experimental paradigm (A) and idealized results (B) from a typical attentional blink paradigm (based on the experiments of Chun & Potter, 1995). The stimuli are presented at fixation at a rate of ten per second. T1 and T2 are digits, and the other stimuli are letters; subjects are required to report the identities of T1 and T2 at the end of the trial. The lag between T1 and T2 is varied, and although T1 accuracy is generally found to be independent of lag, T2 accuracy typically drops significantly at lags 2–4.

called the *attentional blink* because it is analogous to the impairment in accuracy that would result from a T1-triggered eyeblink. This is an interesting example of dual-task interference, and it has generated a great deal of research in recent years. One of the most significant questions is whether the attentional blink reflects a failure to perceive T2 (Raymond, Shapiro, & Arnell, 1992) or whether T2 is perceived but is not stored in a durable form in working memory (Chun & Potter, 1995; Shapiro, Raymond, & Arnell, 1994). Potter (1976) demonstrated that stimuli can be identified more rapidly than they can be stored in working memory, and so it seemed plausible that both T1 and T2 could be identified even though only T1 could be stored in memory when T2 occurred shortly after T1.

To test this hypothesis, Ed Vogel, Kim Shapiro, and I conducted several experiments in which we examined the P1, N1, P3, and N400 components in variations on the attentional blink paradigm (Luck, Vogel, & Shapiro, 1996; Vogel, Luck, & Shapiro, 1998). Here, I will discuss only the most definitive experiment, in which we focused on the N400 component. The N400 component is typically elicited by words that mismatch a previously established semantic context. For example, a large N400 would be elicited by the last word of the sentence, "I opened the dishwasher and pulled out a clean eyebrow," but a small N400 would be elicited if this same sentence ended with the word "plate." The N400 can also be elicited with simple word pairs, such that the second word in PIG-HAT will elicit a large N400 whereas the second word in COAT-HAT will elicit a small N400. The N400 component is well suited for determining whether a word has been identified, because a semantically mismatching word cannot elicit a larger response unless that word has been identified to the point of semantic (or at least lexical) access. Thus, if a given word elicits a larger N400 when it mismatches the semantic context than when it matches the semantic context, this can be taken as strong evidence that the word was identified to a fairly high level. We therefore designed an experiment to determine whether the N400 component would be suppressed for words presented during the attentional blink period. Many previous experiments have examined the effects of semantic mismatch, so we were confident that we could adapt this approach to the attentional blink context (see strategy 2).

Experimental Design   Figure 2.5A illustrates the stimuli and task for this experiment. Each trial began with the presentation of a 1,000-ms "context word" that established a semantic context for that trial. After a 1,000-ms delay, a rapid stream of stimuli was presented at fixation. Each stimulus was seven characters wide. Distractor stimuli were seven-letter sequences of randomly-selected consonants. T1 was a digit that was repeated seven times to form a seven-character stimulus. T2 was a word, presented in red, that

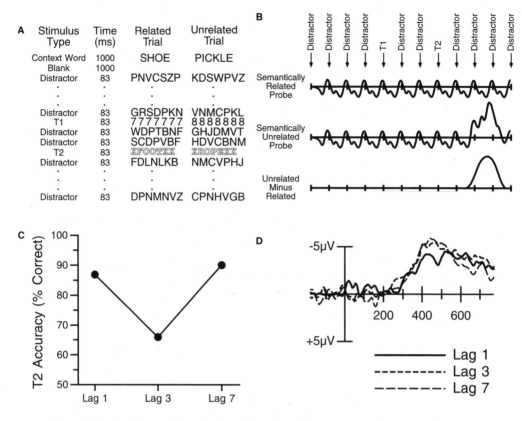

Figure 2.5 Paradigm and results from the study of Vogel, Luck, & Shapiro (1998). (A) Example stimuli. (B) Subtraction method used to overcome the overlap problem. (C) Mean discrimination accuracy for T2 as a function of lag. (D) Grand average ERP difference waveforms from the Cz electrode site, formed by subtracting related-T2 trials from unrelated-T2 trials. Negative is plotted upward.

was either semantically related or semantically unrelated to the context word that had been presented at the beginning of the trial. The T2 word was flanked by Xs, if necessary, to ensure that it was seven characters long. At the end of each trial, the subjects made two responses, one to indicate whether T1 was an odd digit or an even digit, and another to indicate whether T2 was semantically related or unrelated to the context word. Related and unrelated T2 words occurred equally often (as did odd and even T1 digits), and related words were very highly related to the context word (e.g., CAT-DOG). Each T2 word was presented twice for each subject (in widely separated trial blocks), appearing once as a related word and once as an unrelated word. This made it possible to be certain that any differences in the ERPs elicited by related and unrelated words were due to their semantic relationship with the context word rather than any peculiarity of the words themselves (see rule 6 and the Hillyard Principle). The lag between T1 and T2 was either 1, 3, or 7; this restricted set of lags was necessary so that a sufficient number of trials could be obtained at each lag.

As figure 2.5B illustrates the rapid presentation of stimuli in the attentional blink paradigm leads to an overlap problem when recording ERPs. Specifically, the response to a given stimulus lasts for hundreds of milliseconds, overlapping the responses to several of the subsequent stimuli. This makes it difficult to isolate the ERP elicited by T2 from the ERPs elicited by the preceding and subsequent stimuli. To overcome this problem, we computed difference waves (see strategy 4) in which we subtracted the response elicited by T2 when it was semantically related to the context word from the response elicited by T2 when it was an unrelated word (see figure 2.5B). The responses to the other items in the stimulus stream should be essentially identical on related-T2 and unrelated-T2 trials, and this difference wave therefore provides a relatively pure measure of the brain's differential response as a function of the semantic relatedness of T2 relative to the context word. A large difference between related-T2 and unrelated-T2 trials can therefore be used as evidence that the T2 word was identified sufficiently to determine its semantic relationship to the context word.

Results   Panels C and D of figure 2.5 show the results of this experiment. Accuracy for reporting whether T2 was semantically related or unrelated to the context word was highly impaired at lag 3 relative to lags 1 and 7; this is the usual attentional blink pattern. In contrast, the N400 component elicited by T2 was equally large at all three lags. Thus, although the subjects could not accurately report whether T2 was related or unrelated to the context word at lag 3, the N400 wave differentiated between related and unrelated trials, indicating that the brain made this discrimination quite accurately. This result indicates that stimuli are fully identified during the attentional blink, but are not reported accurately because they are not stored in a durable form in working memory.

This conclusion relies on a hidden assumption, namely that the N400 component would be significantly smaller if the perceptual processing of T2 had been impaired at lag 3. This is a problematic assumption, because it is difficult to know what the relationship is between the amplitude of an ERP component and the speed or accuracy of a cognitive process. This brings us to another rule:

Rule 11. Never assume that the amplitude and latency of an ERP component are linearly or even monotonically related to the quality and timing of a cognitive process. This can be tested, but it should not be assumed.

To avoid violating this rule, we conducted a control experiment to determine whether a decrease in the perceptibility of a word would cause a significant decrease in N400 amplitude. In this experiment, we simply added random visual noise of varying intensity to the words. As the intensity of the noise increased, both behavioral accuracy and N400 amplitude declined in a roughly linear manner. Moreover, a change in accuracy that was comparable to the impaired accuracy observed at lag 3 in the main experiment led to a large and statistically significant decline in N400 amplitude. Thus, the absence of a decline in N400 amplitude at lag 3 in the main experiment provides strong evidence that the behavioral errors at lag 3 reflected an impairment that followed word identification.

These results also suggest that a great deal of processing can occur in the absence of awareness, because the brain apparently identified the T2 word at lag 3 even though subjects could not report the word (although it is possible that subjects were briefly aware of the T2 word even though they could not report its semantic relationship moments later).

Larger Issues   I would like to draw attention to three aspects of this attentional blink experiment. First, this experiment used ERPs to monitor a psychological process—word identification—that we suspected was present but could not be observed in the subjects' overt behavior. Many ERP experiments are designed to show correspondences between behavioral results and ERPs, but it is often more interesting to demonstrate an interesting pattern of both similarities and differences. I should note that we ran an additional attentional blink experiment that focused on the P3 wave, and we found that the P3 wave was completely suppressed at lag 3, consistent with the proposal that the attentional blink reflects an impairment in working memory. In this manner, we were able to show a theoretically sensible pattern of similarities and differences between overt behavior and ERPs.

A second notable aspect of this experiment is that we used the N400 component as an index of word identification even though the neural/psychological process that generates the N400 component is probably not directly related to word identification. Because word identification was a necessary antecedent to the presence of an N400 in this experiment, however, we could use the N400 as an indirect index of word identification. Similarly, the P3 wave can be used as an index of "stimulus evaluation time" even though the neural/psychological process that generates the P3 component is probably unrelated to perception. For example, because improbable stimuli elicit a larger P3 than probable stimuli, it is possible to isolate the P3 wave by constructing improbable-minus-probable difference waves. Because the brain cannot produce a larger P3 for an improbable stimulus until that stimulus has been identified and categorized, the timing of the P3 wave in an improbable-minus-probable difference wave can be used as an indirect index of identification and categorization (see Donchin, 1981; Kutas, McCarthy, & Donchin, 1977; McCarthy & Donchin, 1981; Vogel & Luck, 2002).

If you look at ERP experiments that have had a broad impact in cognitive psychology or cognitive neuroscience, you will find that many of them use a given ERP component that is not obviously related to the topic of the experiment. For example, our attentional blink experiment used the language-related N400 component to examine the role of attention in perceptual versus postperceptual processing. Similarly, Dehaene et al. (1998) used the motor-related LRP component to address the possibility of perception without awareness. One of my graduate students, Adam Niese, refers to this as *hijacking* an ERP component, and it leads to an additional strategy of ERP experimental design.

Strategy 7.   Hijack Useful Components from Other Domains
It is useful to pay attention to ERP findings from other subareas of cognitive neuroscience. A component that arises "downstream" from the process of interest may be used to reveal the occurrence of the process of interest.

A third notable aspect of this experiment is that difference waves were used to isolate both the activity elicited by a single stimulus and a specific ERP component elicited by that stimulus (see strategy 4). It is often necessary to present stimuli in such close temporal proximity that the ERPs elicited by each stimulus will overlap each other in time, and difference waves can often be used to circumvent this problem (for additional examples of this approach, see Luck, 1998b; Luck, Fan, & Hillyard, 1993; Luck & Hillyard, 1995). You must use care, however, because you cannot simply assume that the difference wave provides a pure estimate of the size of the component of interest. In the attentional blink experiment, for example, the overall N400 may have been smaller at lag 3 than at lags 1 and 7, even though the difference in N400 amplitude between related and unrelated T2 words was the same at all three lags. In this experiment, the key question was whether the brain could differentiate between related and unrelated T2 words, and the size of the overall N400 was immaterial. In fact, it didn't even matter whether the activity in the unrelated-minus-related difference waves was the N400, the P3, or some other component; no matter what component it was, it demonstrated that the brain could distinguish between related and unrelated words (see strategy 6). Thus, difference waves can be very useful, but they must be interpreted carefully.

Suggestions for Further Reading

There are many examples of clever and thoughtful ERP experimental designs in the literature. The following is a list of a few of my favorites. I would definitely recommend that new ERP researchers study these experiments carefully.

Dehaene, S., Naccache, L., Le Clec'H, G., Koechlin, E., Mueller, M., Dehaene-Lambertz, G., van de Moortele, P. F., & Le Bihan, D. (1998). Imaging unconscious semantic priming. *Nature, 395,* 597–600.

Gehring, W. J., Goss, B., Coles, M. G. H., Meyer, D. E., & Donchin, E. (1993). A neural system for error-detection and compensation. *Psychological Science, 4,* 385–390.

Gratton, G., Coles, M. G. H., Sirevaag, E. J., Eriksen, C. W., & Donchin, E. (1988). Pre- and post-stimulus activation of response channels: A psychophysiological analysis. *Journal of Experimental Psychology: Human Perception and Performance, 14,* 331–344.

Handy, T. C., Solotani, M., & Mangun, G. R. (2001). Perceptual load and visuocortical processing: Event-related potentials reveal sensory-level selection. *Psychological Science, 12*, 213–218.

Magliero, A., Bashore, T. R., Coles, M. G. H., & Donchin, E. (1984). On the dependence of P300 latency on stimulus evaluation processes. *Psychophysiology, 21*, 171–186.

Paller, K. A. (1990). Recall and stem-completion priming have different electrophysiological correlates and are modified differentially by directed forgetting. *Journal of Experimental Psychology: Learning, Memory and Cognition, 16*, 1021–1032.

Van Petten, C., & Kutas, M. (1987). Ambiguous words in context: An event-related potential analysis of the time course of meaning activation. *Journal of Memory & Language, 26*, 188–208.

van Turennout, M., Hagoort, P., & Brown, C. M. (1998). Brain activity during speaking: From syntax to phonology in 40 milliseconds. *Science, 280*, 572–574.

Winkler, I., Kishnerenko, E., Horvath, J., Ceponiene, R., Fellman, V., Huotilainen, M., Naatanen, R., & Sussman, E. (2003). Newborn infants can organize the auditory world. *Proceedings of the National Academy of Sciences, 100*, 11812–11815.

Woldorff, M., & Hillyard, S. A. (1991). Modulation of early auditory processing during selective listening to rapidly presented tones. *Electroencephalography and Clinical Neurophysiology, 79*, 170–191.

Summary of Rules, Principles, and Strategies

Rule 1. Peaks and components are not the same thing. There is nothing special about the point at which the voltage reaches a local maximum.

Rule 2. It is impossible to estimate the time course or peak latency of a latent ERP component by looking at a single ERP waveform—there may be no obvious relationship between the shape of a local part of the waveform and the underlying components.

Rule 3. It is dangerous to compare an experimental effect (i.e., the difference between two ERP waveforms) with the raw ERP waveforms.

Rule 4. Differences in peak amplitude do not necessarily correspond with differences in component size, and differences in peak latency do not necessarily correspond with changes in component timing.

Rule 5. Never assume that an averaged ERP waveform accurately represents the individual waveforms that were averaged together. In particular, the onset and offset times in the averaged waveform will represent the earliest onsets and latest offsets from the individual trials or individual subjects that contribute to the average.

Rule 6. Whenever possible, avoid physical stimulus confounds by using the same physical stimuli across different psychological conditions (the Hillyard Principle). This includes "context" confounds, such as differences in sequential order.

Rule 7. When physical stimulus confounds cannot be avoided, conduct control experiments to assess their plausibility. Never assume that a small physical stimulus difference cannot explain an ERP effect.

Rule 8. Be cautious when comparing averaged ERPs that are based on different numbers of trials.

Rule 9. Be cautious when the presence or timing of motor responses differs between conditions.

Rule 10. Whenever possible, experimental conditions should be varied within trial blocks rather than between trial blocks.

Rule 11. Never assume that the amplitude and latency of an ERP component are linearly or even monotonically related to the quality and timing of a cognitive process. This can be tested, but it should not be assumed.

The Hillyard Principle: Always compare ERPs elicited by the same physical stimuli, varying only the psychological conditions.

Strategy 1. Focus on a specific component.
Strategy 2. Use well-studied experimental manipulations.
Strategy 3. Focus on large components.
Strategy 4. Isolate components with difference waves.
Strategy 5. Focus on components that are easily isolated.
Strategy 6. Use component-independent experimental designs.
Strategy 7. Hijack useful components from other domains.

3 Basic Principles of ERP Recording

This chapter describes how to record clean, artifact-free data. As the ERP technique has matured, there has been a decrease in the amount of discussion in the literature of basic issues such as recording clean data. This is only natural, because a number of laboratories have developed excellent techniques over the years, and these techniques have become a part of the laboratory culture and are passed along as new researchers are trained. However, as time passes, the reasons behind the techniques are often lost, and many new laboratories are using the ERP technique, making it important to revisit the basic technical issues from time to time.

The Importance of Clean Data

Before I begin discussing these issues, I want to discuss why it is important for you to spend considerable time and effort making sure that you are recording the cleanest possible data. The bottom line is that you want to obtain experimental effects that are replicable and statistically significant, and you are unlikely to obtain statistically significant results unless you have low levels of noise in your ERP waveforms. As discussed in chapter 1, the background EEG obscures the ERPs on individual trials, but the ERPs can be isolated from the EEG noise by signal averaging. As you average together more and more trials, the amount of residual EEG noise in the averages will become progressively smaller, so it is crucial to include a sufficient number of trials in your ERP averages. However, increasing the number of trials only works well up to a point, because the effect of averaging on noise is not a direct, linear function of the number of trials; instead, the noise decreases as a function of the square root of the number of trials in the average. As a result, you can't cut the noise in half by doubling the number of trials. In fact, doubling the number of trials decreases the noise only about 30 percent, and you have to include four times as many trials to reduce the noise by 50 percent. Chapter 4 will cover this in more detail.

It should be obvious that you can quadruple the number of trials only so many times before your experiments become absurdly long, so increasing the number of trials is only one part of the solution. The other part is to reduce the noise before it is picked up by the electrodes. Much of the noise in an ERP recording arises not from the EEG, but from non-EEG biological signals such as skin potentials and from electrical noise sources in the environment accidentally picked up during the EEG recording, and it is possible to reduce these sources of noise directly. In fact, if you spend a few days tracking down and eliminating these sources of noise, the resulting improvement in your averaged ERPs could be equivalent to the effects of doubling the number of trials for each subject or the number of subjects in each experiment. This initial effort will be well rewarded in every experiment you conduct.

In addition to tracking down noise sources and eliminating them directly, it is possible to reduce noise by using data processing techniques such as filtering. As chapter 5 will discuss, these techniques are essential in ERP recordings. However, it is important not to depend too much on post-processing techniques to "clean up" a set of ERP data, because these techniques are effective only under limited conditions and because they almost always distort the data in significant ways. This leads us to an important principle that I call *Hansen's Axiom*:

Hansen's Axiom: There is no substitute for good data.

The name of this principle derives from Jon Hansen, who was a scientist and technical guru in Steve Hillyard's lab at UCSD. As he put it in the documentation for a set of artifact rejection procedures:

There is no substitute for good data. It is folly to believe that arti-
fact rejection is going to transform bad data into good data; it can
reject occasional artifactual trials allowing good data to be better.
There is no way that artifact rejection can compensate for a subject
who consistently blinks in response to particular events of interest
or who emits continuous high-amplitude alpha activity. In other
words, data that are consistently noisy or have systematic artifacts
are not likely to be much improved by artifact rejection. (J. C. Han-
sen, unpublished software documentation)

Hansen made this point in the context of artifact rejection, but it
applies broadly to all post-processing procedures that are designed
to clean up the data, ranging from averaging to filtering to independent components analysis. Some post-processing procedures are
essential, but they cannot turn bad data into good data. You will
always save time in the long run by eliminating electrical noise
at the source, by encouraging subjects to minimize bioelectric artifacts, and by designing experiments to produce large effects.

Active and Reference Electrodes

Voltage as a Potential Between Two Sites

Voltage can be thought of as the potential for current to move from
one place to another, and as a result there is really no such thing as
a voltage at a single point (if this fact is not obvious to you, you
should read appendix 1 before continuing). Consider, for example,
a typical twelve-volt automobile battery that has a positive terminal and a negative terminal. The voltage measurement of twelve
volts represents the potential for current to move from the positive
terminal to the negative terminal, and it doesn't make sense to talk
about the voltage at one terminal in isolation. For example, you
could touch one terminal without being shocked (assuming you
weren't touching any other conductors), but if you touch both

terminals you will definitely receive a shock. Similarly, you can never record the voltage at a single scalp electrode. Rather, the EEG is always recorded as a potential for current to pass between two electrodes.

In household electrical systems, a metal stake driven deep into the ground beneath the house serves as an important reference point for electrical devices. The ground literally provides the reference point, and the term *ground* is now used metaphorically in electrical engineering to refer to a common reference point for all voltages in a system. I will therefore use the term *ground* in this general sense, and I will use the term *earth* to mean a stake driven into the ground.

If we measured the electrical potential between an electrode on a subject's scalp and a stake driven into the ground, the voltage would reflect any surplus of electrical charges that had built up in the subject (assuming the subject was not touching a conductor that was connected to earth), and this *static electricity* would obscure any neural signals. We could put an electrode somewhere on the subject's body that was connected to earth, and this would cause any static electricity in the subject to discharge into the earth, eliminating static differences and making it easier to measure changes in neural signals over time. However, it is dangerous to directly connect a subject to earth, because the subject might receive a dangerous shock if touched by an improperly grounded electrical device (such as a button box used for recording responses).

It is possible to create a virtual ground in the amplifier's circuitry that is isolated from earth and connect this ground to a ground electrode somewhere on the subject. You could then record the voltage between a scalp electrode and this ground electrode. However, voltages recorded in this way would still reflect electrical activity at both the scalp electrode and the ground electrode, so it would not provide some sort of absolute measure of electrical activity at the scalp electrode. Moreover, any environmental electrical noise that the amplifier's ground circuit picks up would influence the measured voltage, leading to a great deal of noise in the recording.

To solve the problem of the ground circuit picking up noise, EEG amplification systems use *differential amplifiers*. A differential amplifier uses three electrodes to record activity: an *active* electrode (A) placed at the desired site, a *reference* electrode (R) placed elsewhere on the scalp, and a ground electrode (G) placed at some convenient location on the subject's head or body. The differential amplifier then amplifies the difference between the AG voltage and the RG voltage (AG minus RG). Ambient electrical activity picked up by the amplifier's ground circuit will be the same for the AG and RG voltages and will therefore be eliminated by the subtraction.

It should now be clear that an ERP waveform does not just reflect the electrical properties at the active electrode but instead reflects the difference between the active and reference sites. There is simply no such thing as "the voltage at one electrode site" (because voltage is a potential for charges to move from one place to another). However, there is one way in which it would be meaningful (if slightly imprecise) to talk about voltages at individual sites. Specifically, if one could measure the potential for charges to move from a given point on the scalp to the average of the rest of the surface of the body, then this would be a reasonable way to talk about the voltage at a single electrode site. As I will discuss later in this chapter, points along the body beneath the neck do not matter very much in terms of electrical activity generated by the brain, so we could simplify this by referring to the voltage between one site and the average of the rest of the head. When I later speak of the absolute voltage at a site, I am referring to the potential between that site and the average of the entire head. It is important to keep in mind, however, that absolute voltages are just an idealization; in practice, one begins by recording the potential between two discrete electrode locations.

Originally, researchers assumed that the active electrode was near the active tissue, and the reference electrode was at some distant, electrically neutral site. As the activity near the active electrode changed, researchers assumed that this would influence the voltage at the active site but not at the reference site. When obtaining recordings from several active sites, the same reference electrode is typically used for all of them. For example, you might place electrodes at active locations over frontal, parietal, and occipital cortex and use an earlobe electrode as the reference for all of them.

Figure 3.1 illustrates this, showing the distribution of voltage over the scalp for a single generator dipole (relative to the average of the entire head). The problem with this terminology is that there are no electrically neutral sites on the head (i.e., no sites that are equivalent to the average of the entire scalp), and the voltage recorded between an active site and a so-called reference site will reflect activity at both of the sites. In fact, one of the most important principles that I hope to convey in this chapter is that *an ERP waveform reflects the difference in activity between two sites, not the activity at a single site*. There are some exceptions, of course, which I will describe later in the chapter.

Choosing a Reference Site

Many researchers have tried to minimize activity from the reference electrode by using the most neutral possible reference site, such as the tip of the nose, the chin, the earlobes, the neck, or even the big toe. However, although some of these are useful reference sites, they do not really solve the problem of activity at the reference site. Consider, for example, the tip of the nose. This seems like it ought to be a fairly neutral site because it is fairly far away from the brain. To make the example more extreme, imagine a head with an extremely long nose, like that of Pinocchio. Pinocchio's nose is like a long wire attached to his head, and the voltage will therefore be approximately the same anywhere along this wirelike nose (see figure 3.1B). It doesn't really matter, therefore, whether you place the reference electrode at the tip of the nose or where the nose joins the head—the voltage is equal at both sites,

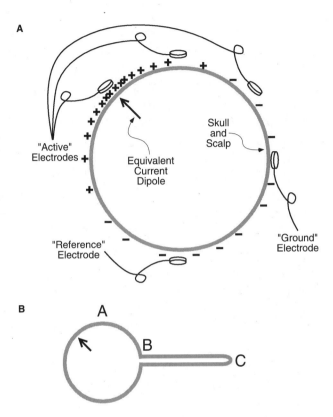

Figure 3.1 Active and reference electrodes. (A) Example of an equivalent current dipole (arrow) inside a spherical head, with the resulting surface voltages (+ and − for positive and negative) on the surface. The recorded voltage will be the difference between the voltage at the active and reference electrodes. (B) Example of the use of a distant reference source. If the active electrode is at point A, it will not matter whether point B or point C is used as the reference, because the voltage at point C will be the same as at point B.

and so the difference between an active site and the reference site will be the same no matter where along the nose the reference electrode is. Because there is no reason to believe that the place where the nose joins the head is more neutral than any other part of the head, the tip of the nose also isn't more neutral than any other part of the head. This is not to say that the tip of the nose is an inappropriate site for a reference electrode. Rather, my point here is that there is no such thing as an electrically neutral reference site, so you must always keep in mind that an ERP waveform reflects contributions from both the active site and the reference site.

What, then, is the best site to use as a reference? There are three main factors involved in choosing a reference. First, given that no site is truly neutral, you might as well choose a site that is convenient and comfortable. The tip of the nose, for example, is a somewhat distracting place for an electrode. Second, you will want to avoid a reference site that is biased toward one hemisphere. For example, if you use the left earlobe as the reference, then the voltage recorded at left hemisphere electrodes will likely be different from the voltage recorded at the right hemisphere. You should avoid this sort of bias. Third, because an ERP waveform for a given active site will look different depending on the choice of the reference site, it is usually a good idea to use the same site for all of your experiments and to use a site that other investigators commonly use. This makes it easier to compare ERP waveforms across experiments and across laboratories.

The most common reference sites in cognitive neuroscience are the earlobes and the mastoid process (the bony protrusion behind each ear). They are close enough to each other that the resulting ERP waveforms should look about the same no matter which you use. The fact that they are commonly used is actually a good reason to continue using them, because this facilitates comparing data across experiments and laboratories. Thus, these sites satisfy the last of the three criteria described in the preceding paragraph. They are also convenient to apply and are not distracting, satisfying the first of the three criteria. In my lab, we use the mastoid rather than the earlobe because we find that an earclip electrode becomes uncomfortable after about an hour and because we find it easier to obtain a good electrical connection from the mastoid (because the skin is not so tough). However, I have heard that one gets a better connection from the earlobe in subjects with questionable personal hygiene. Thus, you should use whichever of these is most convenient for you.

Both the mastoid and the earlobe fail to satisfy the second criterion, because you must pick one side, leading to an imbalance between active electrodes over the left and right hemispheres. The simplest way to avoid this bias is to place electrodes at both the left mastoid (Lm) and the right mastoid (Rm) and then physically connect the wires from the Lm and Rm. This is called a *linked mastoids* reference, and it is not biased toward either hemisphere (you can do the same thing with earlobe electrodes). However, physically linking the wires from Lm and Rm creates a zero-resistance electrical bridge between the hemispheres, which distorts the distribution of voltage over the scalp and reduces any hemispheric asymmetries in the ERPs. To avoid these problems, it is possible to mathematically combine Lm with Rm by using an *average mastoids* reference derivation, which references the active site to the average of Lm and Rm (Nunez, 1981).

There are several ways to do this, and here I'll describe how we do it in my lab. When we record the EEG, we reference all of the scalp sites to Lm, and we also place an electrode on Rm, referenced again to Lm. After recording and averaging the data, we then compute the average mastoids derivation for a given site using the formula $a' = a - (r/2)$, where a' is the desired averaged mastoids waveform for site A, a is the original waveform for site A (with an Lm reference), and r is the original waveform for the Rm site (with an Lm reference).

Let me explain how this formula works. Remember that the voltage at a given electrode site is really the absolute voltage at the active site minus the absolute voltage at the reference site. In the case of an active electrode at site A and an Lm reference, this is $A - Lm$. Similarly, the voltage recorded at Rm is really $Rm - Lm$. The average mastoids reference that we want to compute for site A is the voltage at site A minus the average of the Lm and Rm voltages, or $A - ((Lm + Rm)/2)$. To compute this from the data recorded at A and Rm, we just use some simple algebra:

$a = A - Lm$	Voltage recorded at site A is the absolute voltage at A minus the absolute voltage at Lm.
$r = Rm - Lm$	Voltage recorded at Rm is the absolute voltage at Rm minus the absolute voltage at Lm.
$a' = A - (Lm + Rm)/2$	Average reference derivation at site A is the absolute voltage at site A minus the average of the absolute voltages at Lm and Rm.
$a' = A - Lm/2 - Rm/2$	This is just an algebraic re-organization of the preceding equation.
$a' = A - (Lm - (Lm/2)) - (Rm/2)$	This works because $Lm/2 = Lm - Lm/2$.
$a' = (A - Lm) - ((Rm - Lm)/2)$	This is just an algebraic re-organization of the preceding equation.
$a' = a - (r/2)$	Here we've substituted a for $(A - Lm)$ and r for $(Rm - Lm)$.

In other words, you can compute the voltage corresponding to an average mastoids reference for a given site simply by subtracting half of the voltage recorded from the other mastoid (the same thing can be done with earlobe reference electrodes). All things considered, this is the best reference scheme for the majority of ERP experiments in the area of cognitive neuroscience.

Alternatives to Traditional Reference Sites

There are two additional methods that researchers commonly use to deal with the problem of the reference site. First, imagine that you placed electrodes across the entire surface of the head. By using the average voltage across all of the electrodes as the reference, you could obtain the absolute voltage at each electrode site, and you wouldn't have to worry about the whole reference electrode issue. The mathematics of this would be trivial: the absolute voltage at a given site can be obtained by simply subtracting the average of all of the sites from each individual site, assuming that all sites were recorded with the same reference electrode. Although this would be ideal, it isn't practical for the simple reason that the neck and face get in the way of putting electrodes over about 40 percent of the head.

Some investigators use the average across all of the electrodes as the reference even though they don't have electrodes covering the entire head, and this can lead to some serious misinterpretations (see Desmedt, Chalklin, & Tomberg, 1990). Figure 3.2 illustrates this, showing illustrative waveforms recorded at the Fz, Cz, and Pz electrode sites. The left panel displays the absolute voltage at each site, and a positive-going wave can be seen in condition A relative to condition B at the Fz site. The middle panel shows the voltages that would be obtained using the average of the three electrodes as the reference. When the average of all sites is used as a reference, the average voltage across sites at any given time point is necessarily zero microvolts, so an increase in voltage at one site artificially induces a decrease in voltage at the other sites (the voltages across the entire head also sum to zero, but this is due to physics and not an artificial referencing procedure). As the figure shows, this referencing procedure seriously distorts the waveforms. For example, the short-duration, positive-going peak at around 400 ms at the Fz electrode site becomes a long-duration, negative-going peak when one uses the average reference. This occurs because of the large P3 wave at Pz in the absolute voltage waveforms; to achieve an

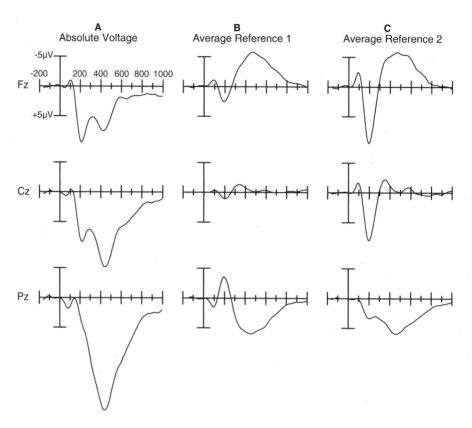

Figure 3.2   Effects of using the average of all sites as the reference. The left column shows the absolute voltage recorded at each of three sites (Fz, Cz, and Pz), and the middle column shows the waveforms that are obtained when using the average of these three sites as the reference. The waveforms are highly distorted when using the average reference. The right column shows the results of using a somewhat larger set of electrodes, covering occipital and temporal electrodes as well as frontal, central, and parietal electrodes. Note that this is not a significant problem when sampling a large proportion of the head's surface (> 60 percent) with a dense set of electrodes (see Dien, 1998). Negative is plotted upward.

average voltage of zero, a large negative voltage has to be added onto the Fz site in the average reference waveforms.

The use of an average reference can lead to serious errors in interpreting ERP data. For example, I have reviewed several manuscripts in which an average-electrodes reference was used and the authors made a great deal out of the finding that an experimental effect was reversed in polarity at some sites relative to others. But this is necessarily the case when using the average across sites as the reference. Thus, it is very dangerous to use the average of all the electrodes as the reference. However, there may be conditions under which the average-electrodes reference might be appropriate. Specifically, Dien (1998) has argued that it is possible to obtain a close approximation of the true average of the entire head by using a large array of electrodes that covers most of the accessible part of the head, and the averaged-electrode reference may therefore be the best approach when enough electrodes are used (assuming that a consensus can be reached about exactly what set of electrodes is necessary to reach an adequate approximation of the true average voltage).

Another approach to this problem is not to rely on voltage measurements, but instead to examine the current flowing out of the head at each point, which does not depend on measuring a difference between an active site and a reference site (unlike voltage, current flow can be legitimately measured at a single point). Specifically, it is possible to convert the voltages measured in standard ERP recordings into current density (sometimes called scalp current density, SCD, or current source density, CSD). Current density is computed for a given electrode site on the basis of the distribution of voltage across the scalp. Technically speaking, one calculates the current density at the scalp by taking the second derivative of the distribution of voltage over the scalp (for a detailed description, see Pernier, Perrin, & Bertrand, 1988). Because ERPs are recorded at a finite set of discrete electrodes, the current density can only be estimated, but given a reasonably dense set of electrodes, the estimate can be quite accurate. More-over, it is not necessary to cover the entire head or even most of the head. If the electrodes are confined to a circumscribed region (e.g., the posterior 50 percent of the head), you can compute accurate estimates of current density within that region, although you shouldn't place much faith in the estimates at the outermost electrodes. Although current density has several advantages, it has one important disadvantage: it is insensitive to dipoles that are deep in the brain and preferentially emphasizes superficial dipoles (this is because the current from a deep source dissipates widely over the entire scalp). Consequently, current density provides a less complete picture of brain activity than traditional voltage measures. In addition, current density is an estimated quantity that is one step removed from the actual data, and it is therefore usually a good idea to examine both the voltage waveforms and the current density waveforms.

Electrical Noise in the Environment

The voltage fluctuations of the scalp EEG are tiny (typically less than 1/100,000th of a volt), and the EEG must be amplified by a factor of 10,000–50,000 before it can be accurately measured. There are many sources of electrical activity in a typical laboratory that are much larger than the EEG, and these large electrical sources can produce small voltage fluctuations in the subject, in the electrodes, and in the wires leading from the subject to the amplifier. These induced voltage changes can be quite considerable when they are amplified along with the EEG (although once the EEG has been amplified, the effects of additional induced voltages are usually insignificant). Although some of the induced electrical noise can be eliminated by filters and other postprocessing techniques, it is always best to eliminate noise at the source (this is a corollary of Hansen's axiom). In this section, I will describe the major sources of electrical noise and discuss strategies for minimizing noise in ERP recordings.

An oscillating voltage in a conductor will induce a small oscillating voltage in nearby conductors, and this is how electrical noise in the environment shows up in the EEG. There are two major sources of oscillating voltages in a typical ERP lab, namely AC line current and video monitors. AC line current consists of sinusoidal oscillations at either 50 Hz or 60 Hz, depending on what part of the world you live in, and this can induce 50- or 60-Hz *line noise* oscillations in your EEG recordings. Video monitors may operate at a refresh rate of anywhere between 50 and 120 Hz (60–75 Hz is typical), but the resulting noise is often spiky rather than sinusoidal. Noise from the video monitor is especially problematic because the stimuli are time-locked to the video refresh, so the noise waveform is the same on every trial and is not reduced by the averaging process.

There are several things that you can do to decrease these sources of noise. The most common approach is to use the amplifier's filters to attenuate the noise. In most cognitive experiments, the ERPs of interest are composed mostly of frequencies under about 30 Hz, so you can filter out everything above 30 Hz (including line noise and video noise) without attenuating the ERPs very much. In addition, many amplifiers have a *line filter* that specifically filters out 50-Hz or 60-Hz noise. However, as chapter 5 will discuss, filters are a form of controlled distortion and their use should be minimized (line filters are especially bad), so filtering alone is not the best way to control noise.

In many laboratories, the subject is placed in an electrically shielded chamber to minimize noise. This can be very effective, but only if there are no noise sources inside the chamber. For example, putting the video monitor inside the chamber creates so much electrical noise that it is hardly worth having a shielded chamber (except in environments with other very large sources of noise). Two approaches are commonly used for solving this problem. First, you can place the video monitor just outside a window in the chamber (but the window must be made of specially treated shielded glass). Second, you can place the monitor inside a

Faraday cage inside of the recording chamber. Figure 3.3A shows a design for a Faraday cage that can be built very easily for a relatively small amount of money (US$200–400). This cage consists of copper screen shielding surrounded by a wooden exterior (with ventilation holes). A shielded piece of glass is placed at the front so that the subject can see the front of the monitor. Shielded glass is available from computer accessory suppliers. There are several different types, and you should get one that has a ground wire coming out of it that you can attach to the copper shielding. You should also get a shielded AC power cord, available at electronics supply companies. A well-shielded Faraday cage can dramatically reduce electrical noise in the EEG, and it is well worth the modest expense.

There may also be other sources of electrical noise inside the chamber. For example, AC lights can be a problem, so you can replace them with DC lights of the type found inside an automobile or recreational vehicle. Basically, you want to make sure there is nothing inside the chamber that might create electrical noise, especially devices powered by AC line voltage. Some of these are obvious, such as lights, but others are not. For example, when I was putting together my current laboratory, I was surprised to find that the cables leading from a stereo amplifier outside the chamber to speakers inside the chamber created significant electrical noise in the EEG—encasing these wires inside shielded conduits eliminated the noise.

Fortunately, there is a fairly easy way to track down sources of electrical noise such as this. As figure 3.3B illustrates, you can create a simulated head out of three resistors, and by connecting this to your amplifier and digitization system, you can see how much noise is present at different locations inside the recording chamber or room. First, place the simulated head where the subject sits and connect the fake head to your amplifier and digitization system using a very high gain (50,000 or 100,000), a fairly high sampling rate (e.g., 1000 Hz), and wide open filter settings (e.g., 0.01–300 Hz). You should see clear signs of electrical noise (including a sinusoi-

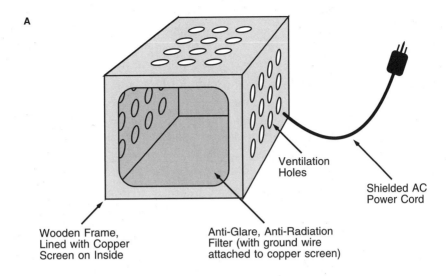

A

Ventilation
Holes

Shielded AC
Power Cord

Wooden Frame,
Lined with Copper
Screen on Inside

Anti-Glare, Anti-Radiation
Filter (with ground wire
attached to copper screen)

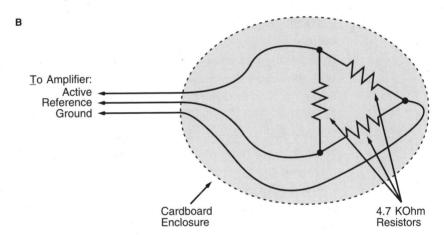

B

To Amplifier:
Active
Reference
Ground

Cardboard
Enclosure

4.7 KOhm
Resistors

Figure 3.3 (A) Faraday cage that can be used to reduce electrical noise from a video monitor. It consists of a plywood box with ventilation holes, completely lined with copper screen (copper sheets may be used on surfaces without ventilation holes). The front of the cage has a glare filter that is covered with a conductive film, which is connected to the copper shielding. A power strip is placed inside the cage (not shown); this is connected to a shielded AC power cord. (B) Simple simulated head for use in finding noise sources. It consists of three 4.7 KOhm resistors (or similar values), connected to form a triangle. Each corner of the triangle is connected to wires leading to the active, reference, and ground inputs of the EEG amplifier. The set of resistors should be enclosed in a nonconductive substance (such as cardboard) for ease of handling.

dal line-frequency oscillation) on your EEG display. Then, turn off every electrical device in or near the recording chamber, including any AC power entering the chamber, such as the power for the fan and lights. This should yield a much cleaner signal on your EEG display (i.e., a nearly flat line). If it doesn't, then you may be picking up noise outside of the chamber, in which case you should shield the cables that bring the EEG from the chamber to your digitization system. Once you have obtained a clean signal, you can turn the electrical devices back on one at a time, noting which ones create noticeable noise. You may also find it useful to place the simulated head near wires and devices that are potential noise sources, which will make it easier to see any noise that they generate.

To identify and eliminate sources of noise in this manner, it is helpful to use software that computes an on-line frequency spectrum of the data from the simulated head. The frequency spectrum will show you the amount of activity in various frequency bands, allowing you to differentiate more easily between line-frequency noise and other sources of noise. This may also allow you to quantify the noise in your data. Once you have minimized the noise level, I would recommend making a printout of the frequency spectrum and posting it in the lab. You can then check the noise level every few months and compare it with this printout to see if the noise level has started to increase (e.g., because the video monitor has been replaced with one that generates more noise).

Electrodes and Impedance

Now that we have discussed the nature of the voltages that are present at the scalp, let's discuss the electrodes that pick up the voltages and deliver them to an amplifier. Basically, a scalp electrode is just a conductive material attached to a wire. In most cases, an electrode is a metal disk that forms an electrical connection with the scalp via a conductive gel. The choice of metal is fairly important, because some metals corrode quickly and loose their conductance. In addition, the circuit formed by the skin, the electrode gel, and the electrode can function as a capacitor that attenuates the transmission of low frequencies (i.e., slow voltage changes).

Until the 1980s, most researchers used silver electrodes covered with a thin coating of silver-chloride (these are typically called *Ag/AgCl* electrodes). These electrodes have many nice properties, but they can be somewhat difficult to maintain. In the 1980s, many investigators started using electrode caps made by Electro-Cap International, which feature tin electrodes that are very easy to maintain. In theory, tin electrodes will tend to attenuate low frequencies more than Ag/AgCl electrodes (Picton, Lins, & Scherg, 1995), but Polich and Lawson found essentially no difference between these two electrode types when common ERP paradigms were tested, even for slow potentials such as the CNV and sustained changes in eye position (Polich & Lawson, 1985). This may reflect the fact that the filtering caused by the electrodes is no more severe than the typical filter settings of an EEG amplifier (Picton, Lins, & Scherg, 1995). Moreover, using high-impedance amplifiers reduces the filtering properties of an electrode. On the other hand, the technology for Ag/AgCl electrodes has improved over the past decade, and now many investigators have switched back to Ag/AgCl. Either tin or Ag/AgCl should be adequate for most purposes, unless you are recording DC potentials.

Because electricity tends to follow the path of least resistance, it is important to ensure that the resistance between the electrode and the scalp is low. Technically speaking, the term *resistance* applies only to an impediment to direct current (DC), in which the voltage does not change over time. When the voltage varies over time (i.e., alternating current or AC), the current can be impeded by *inductance* and *capacitance* as well as resistance; the overall impediment to current flow is called *impedance* (see the appendix for more details). Thus, it is more proper to use the term impedance rather than resistance in the context of ERP recordings, in which the voltage fluctuates over time. Impedance is frequently

Box 3.1   Electrode Naming and Placement Conventions

This section will describe the most common system for placing and naming electrode sites, which was developed in the late 1950s by the International Federation of Clinical Neurophysiology (Jasper, 1958). This system is called the *10/20 system*, because it places electrodes at 10 percent and 20 percent points along lines of latitude and longitude, as illustrated in the figure. The first step in this system is to define an equator, which passes through the nasion (the depression between the eyes at the top of the nose, labeled Nz), the inion (the bump at the back of the head, labeled Iz), and the left and right pre-auricular points (depressions just anterior to the middle of the pinnae, labeled A1 and A2). A longitude line is then drawn between Iz and Nz, and this

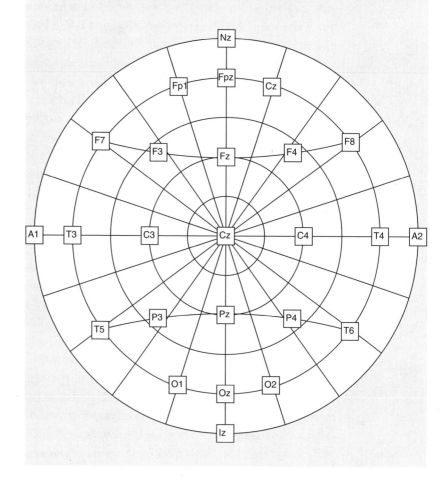

Box 3.1 (continued)

line is then divided into equal sections that are each 10 percent of the length of the line. Additional latitude lines, concentric with the equator, are then placed at these 10 percent points. Most of the electrode sites can then be defined as points that are some multiple of 10 percent or 20 percent along these latitude lines. For example, the F7 electrode is on the left side of the latitude line that is 10 percent from the equator, 30 percent of the distance around the latitude line from the middle. The exceptions to this 10/20 rule are F3 and F4 (halfway between Fz and F7 or F8) and P3 and P4 (halfway between Pz and T5 or T6).

Each electrode name begins with one or two letters to indicate the general region of the electrode (Fp = frontal pole; F = frontal; C = central; P = parietal; O = occipital; and T = temporal). Each electrode name ends with a number or letter indicating distance from the midline, with odd numbers in the left hemisphere and even numbers in the right hemisphere. Larger numbers indicate greater distances from the midline, with locations on the midline labeled with a "z" for zero (because the number 0 looks too much like the letter O).

This scheme has been extended to include many more additional electrode sites (American Encephalographic Society, 1994; Klem et al., 1999), and many individual investigators have developed their own naming schemes. You can use any system you like, as long as it accommodates differences in head sizes and is described in a manner that can be related to the 10/20 system.

denoted by the letter Z and measured in units of Ohms (Ω) or thousands of Ohms (KΩ) (e.g., "the Z for each electrode was less than 5 KΩ").

It is common practice in ERP research to reduce the impedance of the skin to below 5 KΩ before attaching the electrodes; in fact, this practice is so common that I usually neglect to mention it when I write journal articles (although it is definitely worth mentioning). To reduce the impedance, it is necessary to remove the outer layer of dead skin cells that are primarily responsible for the naturally high impedance of the skin. There are two main classes of methods for accomplishing this, and the choice of method depends on the type of electrode being used. If the electrodes are attached to the head individually using some sort of adhesive, you can first clean the skin at each site with an alcohol pad and then rub it with an abrasive paste (it is also possible to do this in one step with alcohol pads that include an abrasive, which I recommend). If you are using an electrode cap, it is not usually possible to abrade the electrode sites before applying the electrodes. Instead, you must insert an abrading implement through the hole in each electrode to abrade the underlying skin. Most laboratories use a blunt needle or the wooden end of a cotton-tipped swab for this. In my lab, we use a sharp, sterile needle. This sounds painful, but it is actually the least painful technique when done properly. You should not stick the needle into the skin, but instead rub it gently along the surface of the skin to displace the top layer of dead skin cells. When done properly, the subject can barely feel the needle. Of course, it is important to thoroughly disinfect the electrodes after each subject, but this is true no matter how one abrades the skin.

Although most investigators know that reducing the impedance is important for obtaining a clean signal, many do not know the precise nature of the problems that high impedance can create. There are really two main types of problems that high impedance can create: (1) decreased common-mode rejection, and (2) increased skin potentials.

First, let's consider common mode rejection. As we discussed near the beginning of this chapter, EEG amplification is accomplished by means of differential amplifiers that amplify the difference between the active-ground voltage and the reference-ground voltage. This subtracts away any electrical noise that is present in the ground (as well as any noise that is equal in the active and reference electrodes) and is essential for obtaining clean EEG recordings. Unfortunately, it is not trivial to produce an amplifier that performs this subtraction perfectly. If one of the two signals in the subtraction is attenuated slightly, then the subtraction does not work and the noise will not be completely subtracted away. The ability of an amplifier to subtract away environmental noise accurately is called *common mode rejection*, and it is usually measured in decibels (dB; an exponential scale in which a doubling of power equals an increase of 3 dB). A good EEG amplifier will have a common mode rejection of at least 70 dB, and preferably over 100 dB. For reasons too complex to explain here, common mode rejection becomes less effective when the impedance is higher, and it can become extremely bad when the impedance is generally high and differs considerably among the active, reference, and ground electrodes. Thus, low electrode impedance helps you avoid picking up environmental noise.

Common mode rejection depends not only on the impedance of the electrodes, but also on the input impedance of the amplifier. An amplifier with a very high input impedance can tolerate higher impedance electrodes while maintaining a good level of common mode rejection. However, high amplifier impedance cannot solve the second problem associated with high electrode impedance, namely *skin potentials*. There is a tonic electrical potential between the surface of the skin and the deep layers of the skin, and this voltage changes whenever the skin's impedance changes. For example, when a subject sweats, the skin's impedance changes and the voltage at the surface changes. Similarly, if the subject moves, shifting the electrode to a slightly different position with a slightly different impedance, the voltage will change. These changes in voltage are called skin potentials, and because they are often very large, they can be a major source of low-frequency noise in ERP recordings. Picton and Hillyard (1972) showed that decreasing the impedance of the skin dramatically reduces skin potentials, and this provides a second compelling reason for decreasing the impedance before recording ERPs.

As computers and electronics have become cheaper and more powerful, ERP researchers have used more and more electrodes in their recordings. In the 1970s, most laboratories used one to three active scalp electrodes, but most laboratories now have the capability to record from at least thirty-two electrodes and many can record from 128 or even 256 electrodes.

Some special problems arise when using dense arrays of electrodes, and I will describe the most important ones here. The biggest drawback to large arrays of electrodes is the time required to decrease the impedance at each site. If it takes an average of one minute to decrease the impedance of each electrode, then a 64-channel electrode cap will require over an hour. A second problem is that the electrode gel inside a given electrode may leak out and cause an electrical bridge with nearby electrodes, distorting the scalp distribution of the electrical potentials. Because the whole point of recording from a dense electrode array is to measure the scalp distribution more accurately, this is a significant problem. The third problem is simply that as the number of electrodes increases, the probability of a bad connection increases and the probability of the experimenter noticing a bad connection decreases.

The geodesic sensor net Tucker and his colleagues developed provides a means of recording from many sites while minimizing some of these problems (Tucker, 1993). The electrodes in this device are basically just sponge-tipped tubes filled with saline, and the electrodes are connected to special high-impedance amplifiers so that no abrasion of the skin is necessary to obtain a reasonable level of common-mode rejection. Consequently, it is possible to attach 128 electrodes in a matter of minutes. Fast application, high-impedance recording systems of various types are now available from several companies. It might appear that these systems solve the problem of the long amount of time required to abrade a large number of sites (and they clearly reduce the risk of disease transmission through the electrodes). However, as discussed above, skin potentials are much larger in high-impedance recordings, and this can be a very considerable source of noise. Moreover, these systems do not solve the problem of a greater likelihood of a bad connection with a larger number of electrodes and the problem of bridging across electrode sites (although some systems can automatically detect bridging).

These high-impedance systems trade speed of application for a degradation in signal quality. Is this a worthwhile tradeoff? The answer to this question depends on how much the signal is degraded and how important it is to record from a large number of electrodes (and to apply them quickly). The issue of signal degradation in these systems has not, to my knowledge, been systematically explored in papers published by independent sources, although there are some publications by people associated with the companies that sell these systems. For example, Tucker's group performed a study of the effects of electrode impedance on noise levels (Ferree et al., 2001), and they found only a modest and statistically insignificant increase in line-frequency (60-Hz) noise as impedance increased. However, they are not unbiased, so they may not have sought out the conditions most likely to produce high noise levels. Moreover, they eliminated skin potentials by filtering out the low frequencies in the signal, making it impossible to assess the effects of high impedance on this very significant source of noise. They argued that 60-Hz noise and skin potentials are not a problem given modern methods, but that has certainly not been my experience!

When making decisions about EEG recording systems, you should keep in mind two important factors. First, as I will discuss in chapter 4, you can offset a decrease in the signal-to-noise ratio by including more trials in your averages, but a very large number of additional trials will be necessary to offset even a modest decrease in the signal-to-noise ratio. For example, you would need to double the number of trials to offset a 30 percent decrease in the signal-to-noise ratio. Thus, you might be able to save an hour by using an easy-to-apply electrode system, but you might need to spend an additional two or three hours collecting data to offset the resulting reduction in signal-to-noise ratio. Before using a fast procedure for applying electrodes, you must be very careful to evaluate the hidden costs that these procedures may entail.

A second key factor is that large numbers of electrodes (> 40) are only occasionally useful in ERP experiments. As discussed in chapters 1 and 7, it is extremely difficult to localize the neural generators of an ERP component solely on the basis of the distribution

Box 3.2 My Personal Perspective on Large Electrode Arrays

I have seen many papers in the last few years in which ERPs were recorded from sixty-four or more electrodes even though the main conclusions of the paper did not necessitate large numbers of electrodes. And in most cases, the data were not very clean. Worse yet, the authors frequently included plots of the waveforms from each site, but given the large number of sites, each waveform was tiny and it was difficult to discern the details of the experimental effects. Because ERPs cannot typically be localized, recording from a large number of channels is usually more of a hindrance than a help. Personally, I would rather have clean data from ten electrodes than noisy data from a thousand electrodes. This brings us again to Hansen's axiom: There is no substitute for good data (and noisy data aren't improved by having lots of channels).

of voltage over the head, so large numbers of electrodes are useful primarily when you will be obtaining some form of converging evidence (e.g., MEG recordings or fMRI scans). Moreover, in the few situations in which large numbers of electrodes are useful, it is also imperative to obtain extremely clean data, which means that common mode rejection should be maximized and skin potentials should be minimized. Thus, there are large costs associated with large electrode arrays, and one should use them only when the benefits clearly outweigh the costs.

Amplifying, Filtering, and Digitizing the Signal

Once the electrodes have picked up the EEG, it must be amplified and then converted from a continuous, analog voltage into a discrete, digital form that a computer can store. Fortunately, these processes are relatively straightforward, although there are a few important issues to consider, such as selecting an amplifier gain and choosing a digitization rate. Although the EEG is amplified before it is digitized, I will discuss the digitization process first because the settings you will use for your amplifier will make more sense once we have discussed the digitization process.

Analog-to-Digital Conversion and High-Pass Filters

A device called an *analog-to-digital converter* (ADC) converts EEG voltage fluctuations into numerical representations. In most EEG digitization systems, the ADC has a resolution of twelve bits. This means that the ADC can code 2^{12} or 4096 different voltage values (intermediate values are simply rounded to the nearest whole number). For example, if the ADC has a range of ± 5 V, a voltage of -5 V would be coded as 0, a voltage of $+5$ V would be coded as 4096, and the intermediate voltages would be coded as 4096 X $((V + 5)/10)$, where V is the voltage level being digitized. Typically, voltages that fall outside the range of the ADC will be coded as 0 for negative values and 4096 for positive values, and you'll obviously want to avoid exceeding this range. Generally, you will want to set the gain on your amplifier so that the range of the ADC is rarely or never exceeded (and, as described in chapter 4, you will want to discard trials that exceed the ADC range). The fact that the EEG is digitized with only twelve bits of resolution does not mean that your data will ultimately be limited to twelve bits of resolution. When you average together many trials, the resolution increases greatly, so twelve bits is sufficient for most cases.

There are two settings on a typical EEG amplifier that will affect whether or not you will exceed the range of the ADC (or the range of the amplifier itself). The first factor is the gain: as you increase the gain, you increase the chance that the EEG will exceed the ADC's range (you don't want to set the gain too low, however, because this will cause a loss of resolution). The second factor is the setting of the high-pass filter. A high-pass filter is a device that attenuates low frequencies and passes high frequencies. High-pass filters are important, because they attenuate the effects of large gradual shifts in voltage due to skin potentials. Even if you use low-impedance recordings, some skin potentials will occur, and these potentials can cause the EEG to drift out of the range of the ADC and amplifier, causing a "flat-line" signal. The higher you set the frequency of the high-pass filter, the less drift will occur. However, as I will discuss further in chapter 5, filters always lead to distortion of the ERPs, and the distortion produced by a high-pass filter almost always becomes worse as you increase the cutoff frequency of the filter. In the 1960s and early 1970s, for example, most ERP experiments used a cutoff frequency of 0.1 Hz, but researchers eventually realized that this cutoff frequency led to a significant reduction in the apparent amplitude of the P3 wave (Duncan-Johnson & Donchin, 1979). As a result, most investigators now use a cutoff of 0.01 Hz, and this is what I would recommend for most experiments. There are, however, some cases in which a higher cutoff frequency would be appropriate. For example, when using children or psychiatric/neurological patients as subjects, voltage drifts are very common, and eliminating these artifacts may be worth some distortion of the ERP waveforms.

In some cases, it is also desirable to record the EEG without using a high-pass filter (these are called direct coupled or DC recordings). Some digitization systems use sixteen-bit ADCs to avoid the problems associated with the lack of a high-pass filter. By using a sixteen-bit ADC, it is possible to decrease the gain by a factor of sixteen, which decreases the likelihood that the EEG will drift out of the range of the ADC. This reduction in gain is equivalent to losing four ADC bits, but this is compensated for by the additional four bits of resolution in the sixteen-bit ADC, yielding the equivalent of twelve bits of resolution. It is also possible to apply a digital high-pass filter to the data at the end of the experiment, which would give you access to both the original unfiltered data and filtered data. However, although this sounds like it would be the best of both worlds, it is difficult in practice to limit a digital filter to very low frequencies. Consequently, it is almost always best to use an amplifier with analog filters that can filter the very low frequencies, reserving DC recordings for the relatively rare cases in which DC data are important for reaching a specific scientific conclusion. I would definitely be wary of buying an amplification system that does not include analog high-pass filters; I have heard several ERP researchers express regret after purchasing DC-only amplifiers.

Discrete EEG Sampling and Low-Pass Filters

The EEG is converted into a voltage at a sequence of discrete time points called samples. The *sampling period* is the amount of time between consecutive samples (e.g., 5 ms) and the *sampling rate* is the number of samples taken per second (e.g., 200 Hz). When many channels are sampled, they are scanned sequentially rather than simultaneously, but the digitization process is so fast that you can think of the channels as being sampled at the same time (unless you are examining extremely high-frequency components, such as brainstem evoked responses). How do you decide what sampling rate to use? To decide, you need to use the *Nyquist theorem*, which states that all of the information in an analog signal such as the EEG can be captured digitally as long as the sampling rate is at least twice as great as the highest frequency in the signal. This theorem also states that you will not only lose information at lower sampling rates, but you will also induce artifactual low frequencies in the digitized data (this is called *aliasing*).

To use the Nyquist theorem, you need to know the frequency content of the signal that you are recording so that you can set your sampling rate to be at least twice as great as the highest frequency in the signal. However, the raw EEG signal may contain noise at arbitrarily high frequencies, so you cannot digitize the raw EEG without risking aliasing. EEG amplifiers therefore contain low-pass filters that attenuate high frequencies and pass low frequencies, and your sampling rate will depend primarily on the cutoff frequency that you select for your low-pass filters. For example, many investigators set the cutoff frequency at 100 Hz and digitize at 200 Hz or faster. As I will discuss in chapter 5, a cutoff frequency of 100 Hz does not completely suppress everything above 100 Hz, so you should use a digitization rate of at least three times the cutoff frequency of the filter. In my laboratory, for example, we usually filter at 80 Hz and digitize at 250 Hz.

The Nyquist theorem gives us a precise means of determining the sampling rate given the filter's cutoff frequency, but how do you decide on the cutoff frequency? Unfortunately, the answer to this question is not so straightforward because, as mentioned above, filters are guaranteed to distort ERP waveforms. In general, the higher the frequency of a low-pass filter, the less distortion it will create. Consequently, you will want to choose a fairly high cutoff frequency for your low-pass filter. However, you don't want your filter frequency to be too high, because this will require a very high digitization rate, leading to huge data files. The best compromise for most cognitive neuroscience experiments is a low-pass cutoff frequency between 30 and 100 Hz and a sampling rate between 100 and 300 Hz. If you are looking at the early sensory responses, you'll want to be at the high end of this range (and even higher if you wish to look at very high-frequency components, such as the brainstem evoked responses). If you are looking only at lower frequency, longer latency components, such as P3 and N400, you can set your cutoff frequency and sampling rate at the low end of this range. Keep in mind, however, that the lower your filter frequency and sampling rate, the less temporal precision you will have.

Amplifier Gain and Calibration

The signal from each electrode is amplified by a separate EEG channel. The gain (amplification factor) that you will use depends on the input range of your analog-to-digital converter. If, for example, your analog-to-digital converter allows an input range of −5 V to +5 V, you will want to set the gain of your amplifier so that its output is near −5 V or +5 V when it has the most extreme possible input values. That is, you will want to use the entire range of the analog-to-digital converter, or else you will not be taking advantage of its full resolution. Most systems work best with a gain somewhere between 1,000 and 50,000 (my lab uses 20,000).

Even if you select the same gain setting for all of the channels of your amplifier, the gains will probably not be exactly the same. It is therefore necessary to calibrate your system. The best way to do this is to pass a voltage of a known size through the system and measure the system's output. For example, if you create a series of 10 µV voltage pulses and run them into your recording system, it may tell you that you have a signal of 9.8 µV on one channel and 10.1 µV on another channel. You can then generate a scaling factor for each channel (computed by dividing the actual value by the measured value), and multiply all of your data by this scaling factor. You can do this multiplication on the EEG data or on the averaged ERP waveforms; the result will be the same.

4   *Averaging, Artifact Rejection, and Artifact Correction*

Because ERPs are embedded in a larger EEG signal, almost all ERP studies rely on some sort of averaging procedure to minimize the EEG noise, and the averaging procedure is typically accompanied by a process that eliminates trials containing artifacts or followed by some procedure to correct for artifacts. These procedures appear to be relatively simple, but there are many important and complex issues lurking below the surface that one must understand before applying them. This chapter will discuss the underlying issues and provide several practical suggestions for averaging and for dealing with artifacts.

The Averaging Process

Basics of Signal Averaging

Figure 4.1 illustrates the traditional approach to signal averaging. First, EEG epochs following a given type of event (usually a stimulus) are extracted from the ongoing EEG. These epochs are aligned with respect to the time-locking event and then simply averaged together in a point-by-point manner. The logic behind this procedure is as follows. The EEG data collected on a single trial is assumed to consist of an ERP waveform plus random noise. The ERP waveform is assumed to be identical on each trial, whereas the noise is assumed to be completely unrelated to the time-locking event. If you could somehow extract just the ERP waveform from the single-trial EEG data, it would look exactly the same on every trial, and averaging together several trials would yield the

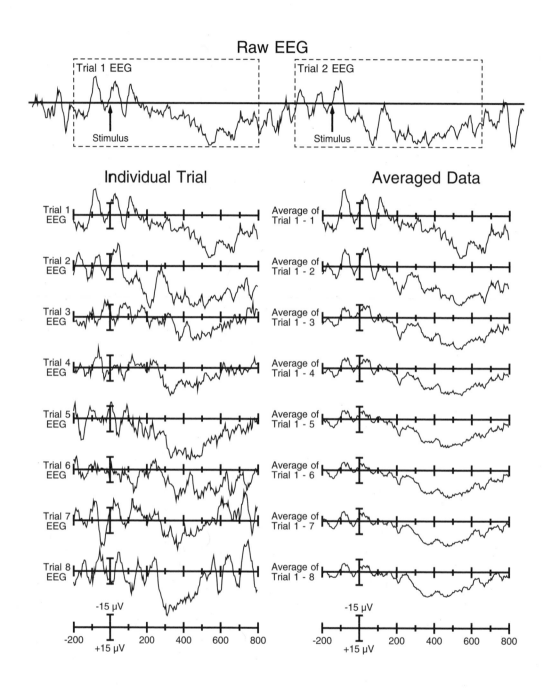

same waveform that was present on the individual trials. In contrast, if you could somehow extract just the noise from the EEG data, it would be random from trial to trial, and the average of a large number of trials would be a flat line at zero microvolts. Thus, when you average together many trials containing both a consistent ERP waveform and random noise, the noise is reduced but the ERP waveform remains.

As you average together more and more trials, the noise remaining in the averaged waveform gets smaller and smaller. Mathematically speaking, if R is the amount of noise on a single trial and N is the number of trials, the size of the noise in an average of the N trials is equal to $(1/\sqrt{N}) \times R$. In other words, the remaining noise in an average decreases as a function of the square root of the number of trials. Moreover, because the signal is assumed to be unaffected by the averaging process, the signal-to-noise (S/N) ratio increases as a function of the square root of the number of trials.

As an example, imagine an experiment in which you are measuring the amplitude of the P3 wave, and the actual amplitude of the P3 wave is 20 μV (if you could measure it without any EEG noise). If the actual noise in the EEG averages 50 μV on a single trial, then the S/N ratio on a single trial will be 20:50, or 0.4 (which is not very good). If you average two trials together, then the S/N ratio will increase by a factor of 1.4 (because $\sqrt{2} = 1.4$). To double the S/N ratio from .4 to .8, it is necessary to average together four trials (because $\sqrt{4} = 2$). To quadruple the S/N ratio from .4 to 1.6, it is necessary to average together sixteen trials (because $\sqrt{16} = 4$). Thus, doubling the S/N ratio requires four times as many trials and quadrupling the S/N ratio requires sixteen times as many trials. This relationship between the number of trials and the S/N

◀ **Figure 4.1**   Example of the application of signal-averaging. The top waveform shows the raw EEG over a period of about 2 seconds, during which time two stimuli were presented. The left column shows segments of EEG for each of several trials, time-locked to stimulus onset. The right column shows the effects of averaging one, two, three, four, five, six, seven, or eight of these EEG segments. Negative is plotted upward.

ratio is rather sobering, because it means that achieving a substantial increase in S/N ratio requires a very large increase in the number of trials. This leads to a very important principle: *It is usually much easier to improve the quality of your data by decreasing sources of noise than by increasing the number of trials.*

To exemplify these principles, figure 4.1 shows the application of signal averaging to a P3 oddball experiment. The top portion of the figure shows the continuous EEG signal, from which the single-trial EEG segments are taken. The left column shows the EEG segments for eight different trials in which an infrequent target was presented. The P3 wave for this subject was quite large, and it can be seen in every trial as a broad positivity in the 300–700 ms latency range. However, there is also quite a bit of variability in the exact shape of the P3 wave, and this is at least partly due to random EEG fluctuations (the P3 itself may also vary from trial to trial). The right column in figure 4.1 shows how averaging together more and more trials minimizes the effects of the random EEG fluctuations. The difference between trial 1 alone and the average of trials 1 and 2 is quite substantial, whereas the difference between the average of trials 1–7 and the average of trials 1–8 is small (even though trial 8 is quite different from the other trials). Note also that the S/N ratio in the average of trials 1–8 is 2.8 times greater than the S/N ratio on the individual trials (because $\sqrt{8} = 2.8$).

The signal-averaging approach is based on several assumptions, the most obvious of which are that (a) the neural activity related to the time-locking event is the same on every trial, and (b) only the EEG noise varies from trial to trial. These assumptions are clearly unrealistic, but violations are not problematic in most cases. For example, if the amplitude of the P2 wave varies from trial to trial, then the P2 wave in the averaged ERP waveform will simply reflect the average amplitude of the P2 wave. Similarly, one could imagine that a P1 wave is present on some trials, a P2 wave is present on other trials, and that both components are never present together on a given single trial. The averaged ERP waveform, however, would contain both a P1 wave and a P2 wave, which would incorrectly imply that the two components are part of the same waveform. However, the conclusions of most ERP experiments do not depend on the assumption that the different parts of the averaged waveform are actually present together on individual trials, so this sort of variability is not usually problematic as long as you always remember that the average is only one possible measure of central tendency.

The Problem of Latency Variability

Although trial-to-trial variability in ERP amplitude is not usually problematic, trial-to-trial variability in latency is sometimes a significant problem. Figure 4.2A illustrates this, showing four individual trials in which a P3-like ERP component occurs at different latencies. The peak amplitude of the average of these trials is much smaller than the peak amplitude on the individual trials. This is particularly problematic when the amount of latency variability differs across experimental conditions. As figure 4.2B shows, a reduction in latency variability causes the peak amplitude of the average to be larger. Thus, if two experimental conditions or groups of subjects differ in the amount of latency variability for some ERP component, they may appear to differ in the amplitude of that component even if the single-trial amplitudes are identical, and this could lead an investigator to incorrectly conclude that there was a difference in amplitude. Worse yet, latency variability can sometimes make it completely impossible to see a given neural response in an averaged waveform (see figure 4.2C). For example, imagine a sinusoidal oscillation that is triggered by a stimulus but varies randomly in phase from trial to trial (which is not just a hypothetical problem—see Gray et al., 1989). Such a response will average to zero and will be essentially invisible in an averaged response.

Figure 4.3 shows a real-world example of latency jitter (from Luck & Hillyard, 1990). In this experiment, we examined the P3 wave during two types of visual search tasks. In one condition

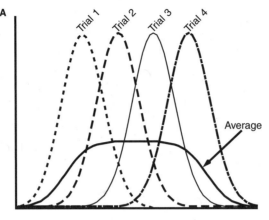

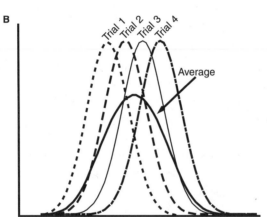

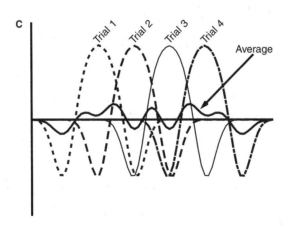

(parallel search), subjects searched for a target with a distinctive visual feature that "popped out" from the display and could be detected immediately no matter how many distractor items were present in the stimulus array. In the other condition (serial search), the target was defined by the absence of a feature, and we expected that in this condition the subjects would search one item at a time until they found the target. In this condition, therefore, we expected reaction time to increase as the number of items in the array (the set size) was increased, whereas we expected no effect of set size in the parallel search condition. This was the pattern of results that we obtained.

We also predicted that both P3 latency and P3 amplitude would increase at the larger set sizes in the serial search condition, but not in the parallel search condition. However, because the order in which the subjects search the arrays is essentially random, we also predicted that there would be more trial-to-trial variability in P3 latency at the larger set sizes in the serial search condition, which made it difficult to measure P3 amplitude and latency. At set size 4, for example, the target could be the first, second, third, or fourth item searched, but at set size 12, the target might be found anywhere between first item and the twelfth item. Consequently, we predicted that P3 latency would be more variable at the larger set sizes in the serial search condition (but not in the parallel search condition, in which the target was detected immediately regardless of the set size).

Figure 4.3B shows the averaged ERP waveforms from this experiment. In the parallel search condition, the P3 wave was relatively large in amplitude and short in duration, and it did not vary much

◀ **Figure 4.2**   Example of the problem of latency variation. Each panel shows four single-trial waveforms, along with the average waveform. The same waveforms are present in panels A and B, but there is greater latency variability in panel A than in panel B, leading to a smaller peak amplitude and broader temporal extent for the average waveform in panel A. Panel C shows that when the single-trial waveforms are not monophasic, but instead have both positive and negative subcomponents, latency variability may lead to cancellation in the averaged waveform.

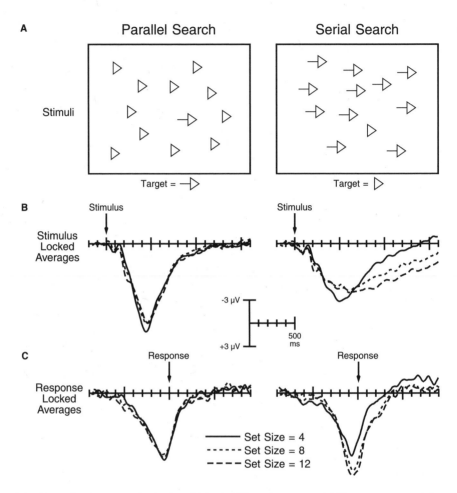

Figure 4.3 Example of an experiment in which significant latency variability was expected for the P3 wave (Luck & Hillyard, 1990). (A) Sample stimuli from the two conditions of the experiment. (B) Stimulus-locked averages. (C) Response-locked averages. Negative is plotted upward.

as a function of set size. In the serial search condition, the P3 had a smaller peak amplitude but was very broad. If you were to measure the amplitude and latency at the peak of the P3 wave in the serial search condition, you might conclude that set size didn't have much of an effect on the P3 in either the serial or parallel search conditions. However, both the amplitude and the latency of the P3 wave were significantly influenced by set size in the serial search condition, although these effects were masked by the latency variability.

I will now describe some techniques that you can use to minimize the effects of latency variability.

Area Measures   In most cases, you can mitigate the reduction in amplitude caused by latency variability simply by using an area amplitude measure rather than a peak amplitude measure. The area under the curve in an average of several trials is always equal to the average of the areas under the curves in each of the individual trials, and an area measure will therefore be completely unaffected by latency variability. In the experiment illustrated in figure 4.3, for example, the area amplitude was significantly greater at larger set sizes, even though peak amplitude was slightly smaller at the larger set sizes. It is also possible to use an area-based measure of latency rather than a peak-based measure. To do this, you simply measure the area under the curve and find the time point that divides this area into equal halves (this is called a *50 percent area latency* measure). When this measure was applied to the data shown in figure 4.3B, the effects of set size on P3 latency were found to be almost identical to the effects on reaction time. Chapter 6 will discuss these area-based measures in greater detail.

Area-based measures are almost always superior to peak-based measures, and insensitivity to latency variability is just one of several reasons why I prefer area-based measures. There are two caveats, however. First, the equivalence between the area in the individual trials and the area in the average is true only when the latency range used to measure the area spans the entire latency range of the component. When there are multiple overlapping ERP components that vary across conditions, it is sometimes necessary to use a relatively restricted measurement window, in which case the area measure is no longer completely insensitive to latency variability (although it's still usually better than a peak amplitude measure). The second caveat is that area measures can misrepresent components that are multiphasic (i.e., components with both positive and negative portions). As figure 4.2C shows, the negative and positive portions of a multiphasic waveform may cancel each other when latency variability is present, and once the data have been averaged there is no way to recover the information that is lost due to this cancellation. Thus, area-based measures are useful for mitigating the effects of latency variability under most conditions, but they are not adequate when there is variability in a multiphasic waveform or when overlapping components preclude the use of a wide measurement window.

Response-Locked Averages   In some cases, variations in the latency of an ERP component are correlated with changes in reaction time, and in these cases latency variability can be corrected by using response-locked averages rather than stimulus-locked averages. In a response-locked average, the response rather than the stimulus is used to align the single-trial EEG segments during the averaging process. Consider, for example, the visual search experiment illustrated in figure 4.3. In the serial search condition, the P3 wave was "smeared out" by latency variability in the stimulus-locked averages, leading to a low peak amplitude and a broad waveform. When response-locked averages were computed, however, the P3 wave in this condition was larger and much more narrowly peaked (see figure 4.3C). In addition, the response-locked averages show that the P3 wave was actually larger at set size 12 than at set size 4, even though the peak amplitude was larger for set size 4 in the stimulus-locked averages. Many studies have used response-locked averages in this manner.

The Woody Filter Technique A third technique for mitigating the effects of latency variability is the *Woody filter* technique (Woody, 1967). The basic approach of this technique is to estimate the latency of the component of interest on individual trials and to use this latency as the time-locking point for averaging. The component is identified on single trials by finding the portion of the single-trial waveform that most closely matches a template of the ERP component. Of course, the success of this technique depends on how well the component of interest can be identified on individual trials, which in turn depends on the S/N ratio of the individual trials and the similarity between the waveshape of the component and the waveshape of the noise.

The Woody filter technique begins with a best-guess template of the component of interest (such as a half cycle of a sine wave) and uses cross-correlations to find the segment of the EEG waveform on each trial that most closely matches the waveshape of the template.[1] The EEG epochs are then aligned with respect to the estimated peak of the component and averaged together. The resulting averaged ERP can then be used as the template for a second iteration of the technique, and additional iterations are performed until little change is observed from one iteration to the next.

The shortcoming of this technique is that the part of the waveform that most closely matches the template on a given trial may not always be the actual component of interest, resulting in an averaged waveform that does not accurately reflect the amplitude and latency of the component of interest (see Wastell, 1977). Moreover, this does not simply add random noise to the averages; instead, it tends to make the averages from each different experimental condition more similar to the template and therefore more similar to each other (this is basically just regression toward the mean). Thus, this technique is useful only when the component of interest is relatively large and dissimilar to the EEG noise. For example, the P1 wave is small and is similar in shape to spontaneous alpha waves in the EEG, and the template would be more closely matched by the noise than by the actual single-trial P1 wave on many trials. The P3 component, in contrast, is relatively large and differs in waveshape from common EEG patterns, and the template-matching procedure is therefore more likely to find the actual P3 wave on single trials.

However, even when one is examining a large component such as the P3 wave, Woody filtering works best when the latency variability is only moderate; when the variability is great, a very wide window must be searched on the individual trials, leading to more opportunities for a noise deflection to match the template better than the component of interest. For example, I tried to apply the Woody filter technique to the visual search experiment shown in figure 4.3, but it didn't work very well. The P3 wave in this experiment could peak anywhere between 400 and 1,400 milliseconds poststimulus, and given this broad search window, the algorithm frequently located a portion of the waveform that matched the search template fairly well but did not correspond to the actual P3 peak. As a result, the averages looked very much like the search template and were highly similar across conditions.

One should note that the major difficulty with the Woody filter technique lies in identifying the component of interest on single trials, and any factors that improve this process will lead to a more accurate adjustment of the averages. For example, the scalp distribution of the component could be specified in addition to the component's waveshape, which would make it possible to reject spurious EEG deflections that may have the correct waveshape but have an incorrect scalp distribution (see Brandeis et al., 1992).

Time-Locked Spectral Averaging   The final technique considered here is a means of extracting oscillatory responses that have random phase (onset time) with respect to the time-locking event. As discussed at the beginning of this section, oscillations that vary in phase will be lost in a conventional average, but one can see these oscillations using techniques that measure the amplitudes of the oscillations on single trials and then average these amplitude measures across trials. The techniques for measuring single-trial oscillation amplitudes rely on variants of a mathematical procedure called the *Fourier transform*. As I will discuss more fully in chapter 5, the Fourier transform converts a waveform into a set of sine waves of different frequencies, phases, and amplitudes. For example, if you were to apply the Fourier transform to a 1-second EEG epoch, you would be able to determine the amount of activity at 10 Hz, at 15 Hz, at 20 Hz, or almost any frequency. In this manner, you could compute the amplitude at each frequency for a single trial, and you could then average these amplitude measures across trials. Moreover, because the amplitude is measured independently of the phase, it wouldn't matter if the phase (i.e., the latency) of the oscillations varied across trials.

Although this is a useful approach, it completely discards the temporal information of the ERP technique, because the amplitude measured on a given trial is the amplitude for the entire time period. Temporal information can be retained, however, by using *moving window* techniques, such as those developed by Makeig (1993) and by Tallon-Baudry and colleagues (1996). These techniques extract a brief window of EEG from the beginning of the trial (e.g., the first 100 ms of EEG). The Fourier transform is then applied to this window to provide a quantification of the amplitude at each frequency during that relatively brief time range. The window is then moved over slightly (e.g., by 10 ms), and another Fourier transform is applied. In this manner, it is possible to compute Fourier transforms for every point in the EEG, although the values at a given time point actually represent the frequencies over a period of time (e.g., a 100-ms period). The Fourier transforms at a given time point are then averaged across trials just as the EEG amplitude would be averaged across trials in a conventional average. This is called *time-locked spectral averaging* because time-locked averages are computed for spectral (i.e., frequency) information.

Figure 4.4A[2] shows an example of this technique, presenting time-locked spectral averages from the study of Tallon-Baudry et al. (1996), who were interested in the 40-Hz oscillations elicited

A) Time X Frequency Activity on Single Trials

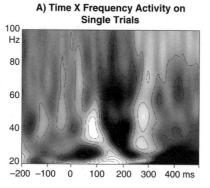

B) Time X Frequency Activity from Averaged ERPs

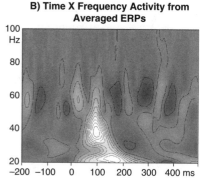

Figure 4.4 Example of time-locked spectral averaging. In panel A, the frequency transformation was applied to the individual trials and the transformed data were then averaged. This plot therefore includes activity that was not phase-locked to stimulus onset as well as phase-locked activity. In panel B, the transformation was applied after the waveforms had been averaged together. This plot therefore includes only activity that was phase-locked to the stimulus, because random-phase activity is eliminated by the ERP averaging process. (Adapted with permission from Tallon-Baudry et al., 1996. © 1996 Society for Neuroscience.)

by visual stimuli. The X-axis in this plot is time, just as in a traditional ERP average. The Y-axis, however, is frequency, and the gray-scale level indicates the power that was present at each frequency at each time point. A band of activity between 40 and 50 Hz can be seen at approximately 100 ms poststimulus, and a somewhat weaker band of activity between 30 and 60 Hz can be seen at approximately 300 ms poststimulus. Activity can also be seen in the 20-Hz range from about 100 to 200 ms poststimulus.

The crucial aspect of this approach is that these bands of activity can be seen whether or not the oscillations vary in phase from trial to trial, whereas random-phase activity is completely lost in a traditional average. Time-locked spectral averaging thus provides a very useful technique for examining random-phase oscillations. However, it is very easy to draw an incorrect conclusion from data such as those shown in figure 4.4A, namely that the activity really consists of oscillations. As I will discuss fully in chapter 5, a brief monophasic ERP deflection contains activity at a variety of frequencies, and the presence of activity in a given frequency band does not entail the existence of a true oscillation (i.e., an oscillation with multiple positive and negative deflections). For example, figure 4.4B shows the time × frequency transformation of the traditional ERP averages from the study of Tallon-Baudry et al. (1996), and the 40–50 Hz activity at 100 ms poststimulus can be seen in this plot just as in the single-trial data. In other words, this activity was a part of the traditional ERP and was not a random-phase oscillation. Thus, to draw conclusions about random-phase oscillations, it is necessary to apply the time × frequency transformation to the averaged ERPs as well as to the single-trial EEG data.

Overlap from Preceding and Subsequent Stimuli

Overlapping ERPs from previous and subsequent stimuli will distort averaged ERP waveforms in ways that are sometimes subtle and sometimes obvious, and it is important to understand how this arises and when it might lead you to misinterpret your data (for a detailed treatment of this issue, see Woldorff, 1988). Overlap arises when the response to the previous stimulus has not ended before the baseline period prior to the current stimulus or when the subsequent stimulus is presented before the ERP response to the current stimulus has terminated. This problem is particularly acute when stimuli are presented rapidly (e.g., 1 second or less between stimulus onsets). However, ERP waveforms can last for several seconds, and overlap can significantly distort your data even at long interstimulus intervals.

Figure 4.5 illustrates the overlap problem for a thought experiment in which a stimulus is presented every 300–500 ms. Panel A shows the actual waveform elicited by a given stimulus without any overlap. Note that the prestimulus period is flat and that the waveform falls to zero at 1000 ms poststimulus. Panel B shows the waveforms that the same stimulus would have produced if

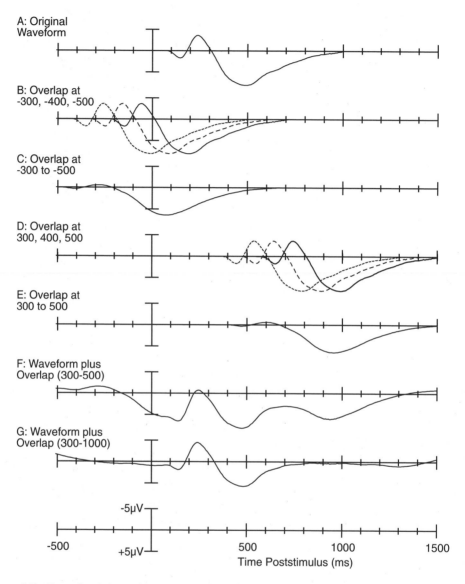

Figure 4.5   Example of the problem of overlapping waveforms. (A) An example ERP waveform. (B) Waveforms produced by the previous stimulus when it appears 300, 400, or 500 ms prior to the current stimulus. (C) Average waveform produced by the previous stimulus when it appears at random times between 300 and 500 ms prior to the current stimulus. (D) Waveforms produced by the subsequent stimulus when it appears

it appeared 300, 400, or 500 ms prior to the current stimulus; these are just the original waveform shifted to the left by various amounts. Panel C shows the average of a large number of previous waveforms, elicited by stimuli happening randomly and equiprobably between 300 and 500 ms prior to the current stimulus. This is the average overlap from the preceding stimuli. Panel D shows the responses elicited by the subsequent stimulus at 300, 400, or 500 ms, and panel E shows the overlap that would occur with stimuli occurring randomly 300–500 ms after the current stimulus.

Note that the jittered timing in this thought experiment leads to a "smearing" of the averaged waveform for the overlapping stimuli. That is, the relatively sharp positive and negative peaks at the beginning of the original waveform are mostly (but not entirely) eliminated in the overlapping waveform. The effect of temporal jitter between stimuli is equivalent to filtering out the high frequencies from the original waveform. As the range of time delays between the stimuli becomes wider and wider, the jitter reduces lower and lower frequencies from the overlap. However, even with a broad jitter, some low-frequency overlap will still occur (chapter 5 will describe a set of mathematical formalizations that you can use to understand in detail the filtering properties of temporal jitter).

Panel F shows the sum of the original waveform and the overlapping waveforms. This sum is exactly what you would obtain if you simply averaged together all of the stimuli in this thought experiment. The distortion due to overlap is quite severe. First, the prestimulus baseline is completely distorted, and this leads the initial positive component to seem much larger than it really is (compare the first positive component in panel G to the original waveform

◀ **Figure 4.5** (continued)
300, 400, or 500 ms after the current stimulus. (E) Average waveform produced by the subsequent stimulus when it appears at random times between 300 and 500 ms after the current stimulus. (F) Sum of the original waveform and the overlapping waveforms from (C) and (E). (G) Sum of the original waveform and the overlapping waveforms that would be produced if the interval between stimuli was increased to 300–1000 ms. Negative is plotted upward.

in panel A). Second, the first positive component appears to start before time zero (which is always a good indication that something is amiss). Third, there is a late positive peak that is completely artifactual.

Overlap is particularly problematic when it differs between experimental conditions. Imagine, for example, that the same target stimulus is presented in condition A and in condition B, and the preceding stimulus elicits a large P3 wave in condition A and a small P3 wave in condition B. This difference in the preceding stimuli may influence the prestimulus baseline period, and this in turn will influence the apparent amplitude of the P3 elicited by the target stimuli on the current trial. Note that this can happen even if the ERP response to the preceding stimulus has returned to baseline before the P3 wave on the current trial: if the baseline is affected, then the whole waveform will be affected. This can also have a big impact on attempts to localize the generator of an ERP component, because the scalp distribution of the overlapping activity in the prestimulus period will be subtracted away from the scalp distribution of the component that you are trying to localize, distorting the apparent scalp distribution of the component.

There are some steps that you can take to minimize the effects of overlap. The first and most important step is to think carefully about exactly what pattern the overlap will take and how it might differ across conditions (for an example, see Woldorff & Hillyard, 1991). The key is to think about how the task instructions may change the response to the preceding stimulus across conditions, even if it is the same physical stimulus. You may even want to simulate the overlap, which isn't conceptually difficult (for details, see Woldorff, 1988). It's computationally trivial to do this in a programming environment such as MATLAB, and it can even be done fairly easily in Excel.

A second approach is to use the broadest possible range of time delays between stimuli. Panel G of figure 4.5 shows what happens if the jitter in our thought experiment is expanded to 300–1000 ms. There is still some overlap, but it is much smaller.

A third approach is to use high-pass filters to filter out any remaining overlap. As mentioned earlier in this section, jittering the interstimulus interval is equivalent to filtering out the high frequencies from the overlap, and the remaining low frequencies can be filtered offline with a high-pass filter. High-pass filtering can distort your waveforms in other ways, however, so you must do this cautiously (see chapter 5 for details).

A fourth approach is to design the experiment in a way that allows you to directly measure and subtract away the overlap. In our thought experiment, for example, we could include occasional trials on which no stimulus was presented and create averaged waveforms time-locked to the time at which the stimulus would ordinarily have occurred. These waveforms will contain only the overlap, and these overlap waveforms can be subtracted from the averaged waveforms that contained the response to an actual stimulus along with the overlap. This requires some careful thought, however, because the subject might notice the omission of a stimulus triggering an ERP (see, e.g., Picton, Hillyard, & Galambos, 1974). I have frequently used this approach in my own research (see, in particular, Luck, 1998b; Vogel, Luck, & Shapiro, 1998).

A fifth approach is to estimate the overlap from the data you have collected and subtract the estimated overlap from the averaged ERP waveforms. Woldorff (1988) has developed a technique called the *ADJAR* (adjacent response) filter that has been used for this purpose in a number of experiments (e.g., Hopfinger & Mangun, 1998; Luck et al., 1994; Woldorff & Hillyard, 1991).

Transient and Steady-State ERPs

If stimuli are presented at a constant rate rather than a variable rate, the overlap from the preceding and subsequent stimuli is fully present in the averaged ERP waveforms. In fact, the preceding and subsequent stimuli are perfectly time-locked to the current stimulus, so it makes sense that they would appear in the averaged ERP waveforms. Most cognitive ERP experiments therefore use some jitter in the interstimulus interval unless the interstimulus interval is fairly long.

It is sometimes possible to make overlap into a virtue. Specifically, if a series of identical stimuli are presented at a fast, regular rate (e.g., eight stimuli per second), the system will stop producing complex *transient* responses and enter into a *steady state*, in which the system resonates at the stimulus rate (and multiples thereof). Typically, steady state responses will look like two summed sine waves, one at the stimulation frequency and one at twice the stimulation frequency.

Figure 4.6 shows an example of a steady state response. The upper left portion of the figure shows the transient response obtained when the on-off cycle of a visual stimulus repeats twice per second. When the stimulation rate is increased to 6 cycles per second, it is still possible to see some distinct peaks, but the overall waveform now appears to repeat continuously, with no clear beginning or end. As the stimulation rate is increased to 12 and then 20

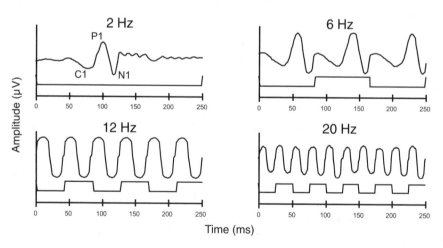

Figure 4.6   Transient response to a stimulus presented at a rate of two on-off cycles per second and steady-state response to a stimulus presented at 6, 12, or 20 Hz. Positive is plotted upward. (Adapted with permission from Di Russo, Teder-Sälejärvi, & Hillyard, 2003. © 2003 Academic Press.) Thanks to Francesco Di Russo for providing an electronic version of this figure.

cycles per second, the response is predominantly a sine wave at the stimulation frequency (with a small, hard-to-see component at twice the stimulation frequency).

This steady-state response can be summarized by four numbers, the amplitude (size) and phase (temporal shift) of each of the two sine waves. This is a lot simpler than a complex transient response with a separate amplitude value at each point in time. As a result, steady-state ERPs are widely used in the study of sensory systems and in the diagnosis of sensory disorders.

Steady-state ERPs have a significant shortcoming, however, which is that they do not provide very precise temporal information. For example, if stimuli are presented every 150 ms, the voltage measured at 130 ms after the onset of one stimulus consists of the sum of the response to the current stimulus at 130 ms, the response to the previous stimulus at 280 ms, the response to the stimulus before that at 430 ms, and so on. Because steady-state ERPs lack the high temporal resolution of transient ERPs, they are used only rarely in cognitive studies (for a review of some recent cognitive steady-state studies, see Hopfinger, Luck, & Hillyard, 2004).

Artifact Rejection and Correction

Now that we have considered the averaging process, we will move on to the artifact rejection procedures that typically accompany it.

There are several types of artifacts that can contaminate EEG recordings, including blinks, eye movements, muscle activity, and skin potentials. These artifacts can be problematic in two ways. First, they are typically very large compared to the ERP signals and may greatly decrease the S/N ratio of the averaged ERP waveform. Second, some types of artifacts may be systematic rather than random, occurring in some conditions more than others and being at least loosely time-locked to the stimulus so that the averaging process does not eliminate them. Such artifacts may lead to erroneous conclusions about the effects of an experimental manipulation.

For example, some stimuli may be more likely to elicit blinks than others, which could lead to differences in amplitude in the averaged ERP waveforms.

There are two main classes of techniques for eliminating the deleterious effects of artifacts. First, it is possible to detect large artifacts in the single-trial EEG epochs and simply exclude contaminated trials from the averaged ERP waveforms (this is called *artifact rejection*). Alternatively, it is sometimes possible to estimate the influence of the artifacts on the ERPs and use correction procedures to subtract away the estimated contribution of the artifacts (this is called *artifact correction*). In this section, I will discuss both approaches. However, I would first like to make a point that should be obvious but is often overlooked. Specifically, it is always better to minimize the occurrence of artifacts rather than to rely heavily on rejection or correction procedures. This is really just a special case of Hansen's Axiom: there is no substitute for good data. In other words, time spent eliminating artifacts at the source will be well rewarded. This section will therefore also include hints for reducing the occurrence of artifacts.

The General Artifact Rejection Process

Before I get into the details of how to detect specific types of artifacts, I would like to provide a general framework for conceptualizing the artifact rejection process.[3] Detecting artifacts is, in essence, a signal detection problem, in which the artifact is treated as the to-be-detected signal. As an example, imagine that you have lost a valuable ring on a beach, and you have rented a metal detector to help you find it. The metal detector has a continuously variable output that tells you the extent to which there is evidence of nearby metal, but this output is quite variable due to random fluctuations in the mineral content of the sand. If you started digging in the sand any time there was a hint of nearby metal, you would make very slow progress because you would start digging every few feet. However, if you only started digging when the metal detector's output was very high, you might miss the ring altogether because it is small and doesn't create a large change in the detector's output. Thus, if you dig only when the detector's output reaches a very high level, you will probably pass right over the top of the ring, but if you dig whenever the output exceeds some small level, you will frequently be digging in vain.

The key aspects of this example are as follows. You are trying to detect something that is either there or not (the ring) based on a noisy, continuously variable signal (the metal detector's output). You select a threshold value, and if the signal exceeds that value, you make a response (digging). In this context, we can define four outcomes for each patch of sand: (1) a hit occurs when the sought-after object is present, the signal exceeds the threshold, and you respond (i.e., the metal detector's output exceeds a certain value, you dig, and you find the ring); (2) a miss occurs when the object is present, the signal fails to exceed the threshold, and you don't respond; (3) a false alarm occurs when the object is absent, the signal exceeds the threshold due to random variation, and you respond; (4) a correct rejection occurs when the object is absent, the signal doesn't exceed the threshold, and you don't respond. Hits and correct rejections are both correct responses, and misses and false alarms are both errors. Importantly, you can increase the number of hits by choosing a lower threshold (i.e., digging when the metal detector's output is fairly low), but this will also lead to an increase in the number of false alarms. The only way to increase the hit rate without increasing the false alarm rate is to get a better metal detector with an output that better differentiates between the presence or absence of small metal objects.

Now imagine that you are trying to detect blinks in a noisy EEG signal. When a subject blinks, the movement of the eyelids across the eyeball creates a voltage deflection, and it is possible to assess the presence or absence of a blink by measuring the size of the largest voltage deflection within a given segment of EEG (just like assessing the presence or absence of the ring by examining the output of the metal detector). If the voltage deflection exceeds a certain threshold level, you conclude that the subject blinked and you discard that trial; if the threshold is not exceeded, you conclude that the subject did not blink and you include that trial in the averaged ERP waveform. If you set a low threshold and reject any trials that have even a small voltage deflection, you will eliminate all of the trials with blinks, but you will also discard many blink-free trials, reducing the signal-to-noise ratio of the averaged ERP waveform. If you set a high threshold and reject only trials with very large voltage deflections, you will have more trials in your averages, but some of those trials may contain blinks that failed to exceed your threshold. Thus, simply changing the threshold cannot increase the rejection of true artifacts without also increasing the rejection of artifact-free trials. However, just as you can do a better job of finding a ring in the sand by using a better metal detector, you can do a better job of rejecting artifacts by using a better procedure for measuring artifacts.

Choosing an Artifact Measure

Many software systems assess blinks by measuring the maximal voltage in the EOG signal on a given trial and rejecting trials in which this maximal voltage exceeds a threshold, such as ± 75 μV. However, the peak amplitude in the EOG channel is a very poor measure of blink artifacts, because variations in the baseline voltage can bring the EOG signal far enough away from zero that small noise deflections will sometimes cause the voltage to exceed the threshold voltage, causing false alarms. Variations in baseline voltage can also bring the EOG signal *away* from the threshold voltage so that a true blink no longer exceeds the threshold, leading to misses.

Figure 4.7 illustrates these problems. Panel A of the figure shows an EOG recording with a blink that exceeds the 75 μV threshold and would be correctly rejected. Panel B shows an epoch in which a blink is clearly present, but because the baseline has shifted, the 75 μV threshold is not exceeded (a miss). Panel C

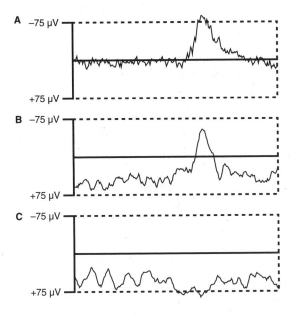

Figure 4.7   Example of the use of an absolute voltage threshold for artifact rejection. Each waveform shows single-trial activity recorded from an EOG electrode under the left eye, with a right-mastoid reference. In panel A, a blink is present and exceeds the threshold. In panel B, a blink is again present, but because the voltage had drifted downward, the threshold is not exceeded. In panel C, no blink is present, but because of a downward drift in the signal, the threshold is exceeded. Negative is plotted upward.

shows an epoch in which no blink was present, but a shift in baseline causes the 75 μV threshold to be exceeded by simple noise (a false alarm).

An alternative approach is to measure the difference between the minimum and maximum voltages within an EOG epoch and to compare this peak-to-peak voltage with some threshold voltage. This peak-to-peak measure is less distorted by slow changes in baseline voltage, reducing the impact of this possible source of misses and false alarms, and increasing the sensitivity of the artifact rejection process. Thus, it is possible to increase the rejection of trials with artifacts without increasing the rejection of artifact-free trials by choosing a measure that can accurately distinguish between trials with and without artifacts. Later in this chapter I'll offer some suggestions for good measures.

The choice of the most sensitive measure will depend on what type of artifact you are measuring. For example, although peak-to-peak amplitude is a sensitive measure of blinks, it is not a very sensitive measure of the presence of alpha waves, which are frequently no larger than the background EEG activity. A more sensitive measure of alpha waves would be, for example, the amount of power in the 10-Hz frequency range, which would be high on trials contaminated by alpha waves and low for uncontaminated trials. Thus, a good artifact rejection system should allow the use of different measures for different types of artifacts.

It is important to note that a measure such as peak-to-peak amplitude is not really a measure of blinking, but is simply a numeric value that you can calculate from the data and use to differentiate probabilistically between trials with and without blinks. The artifact rejection process can thus be conceptualized in the general case as a two-step process, in which a "function" is applied to the data to compute a specific value and then this value is compared to the threshold. You can use different functions and different criteria in different cases, depending on the nature of the artifact that you are trying to detect.

Some investigators visually inspect the EEG on each trial to determine which trials contain artifacts, but this process is conceptually identical to the procedure that I just outlined. The only difference is that it uses the experimenter's visual system instead of a computer algorithm to determine the extent to which an artifact appears to be present and uses an informal, internal threshold to determine which trials to reject. The advantage of this approach is that the human visual system can be trained to do an excellent job of differentiating between real artifacts and normal EEG noise. However, a well-designed computer algorithm may be just as sensitive, if not more so. And computer algorithms have the advantages of being fast and not being prone to bias. Thus, it is usually best to use a good automated artifact rejection system rather than spending hours trying to identify artifacts by eye.

Choosing a Rejection Threshold

Once you have chosen an appropriate measure of an artifact, you must choose the threshold that will be used to determine whether to reject an individual trial. One possibility is to pick a threshold on the basis of experience and use this value for all subjects. For example, you may decide to reject all trials with a peak-to-peak EOG amplitude of 50 µV or higher. However, there is often significant variability across subjects in the size and shape of the voltage deflections a given type of artifact produces and in the characteristics of the EEG in which these voltage deflections are embedded, so a one-size-fits-all approach is therefore not optimal. Instead, it is usually best to tailor the artifact rejection process for each individual subject.

There is at least one exception to this rule, however: experiments that use different subject groups and in which the artifact rejection process could lead to some sort of bias. For example, it might not be appropriate to use different artifact rejection criteria for different subjects in a study that compared schizophrenia patients with normal controls, because any differences in the resulting averaged ERP waveforms could reflect a difference in artifact rejection rather than a real difference in the ERPs. However, using the same criteria for all subjects could also be problematic in such a study if, for example, one group had smaller blink amplitudes than another, resulting in more contamination from artifacts that escaped rejection. The best compromise in between-subject studies is probably to set the criteria individually for each subject, but to be blind to the subject's condition when the criteria are set. In addition, it would be worthwhile to determine whether the results are any different when using the same threshold for all subjects compared to when tailoring the threshold for each subject—if the results are the same, then the threshold is not causing any bias in the results.

If the threshold is set individually for each subject, the settings are usually based on visual inspection of a portion of the raw EEG. This can be accomplished by the following sequence of steps. First, select an initial threshold for a given subject as a starting point (usually on the basis of experience with prior subjects). Then apply this threshold to a set of individual trials and visually assess whether trials with real artifacts are not being rejected or if trials without real artifacts are being rejected. Of course, this requires that you are able to determine the presence or absence of artifacts by visual inspection. In most cases, this is fairly straightforward, and the next section provides some hints. After the initial threshold has been tested on some data, it can be adjusted and retested until it rejects all of the trials that clearly have artifacts without rejecting too many artifact-free trials (as assessed visually). Some types of artifacts also leave a distinctive "signature" in the averaged waveforms, so it is also possible to evaluate whether the threshold adequately rejected trials with artifacts after you have averaged the data.

It can also be useful to ask the subject to make some blinks and eye movements at the beginning of the session so that you can easily see what that subject's artifacts look like.

Detecting and Rejecting Specific Types of Artifacts

In this section, I will discuss several common types of artifacts and provide suggestions for reducing their occurrence and for detecting and rejecting them when they do occur.

Blinks   Within each eye, there is an electrical gradient with positive at the front of the eye and negative at the back of the eye, and the voltage deflections recorded near the eye are primarily caused by the movement of the eyelids across the eyes, which modulates the conduction of the electrical potentials of the eyes to the surrounding regions. Figure 4.8 shows the typical waveshape of the eyeblink response at a location below the eyes (labeled VEOG) and

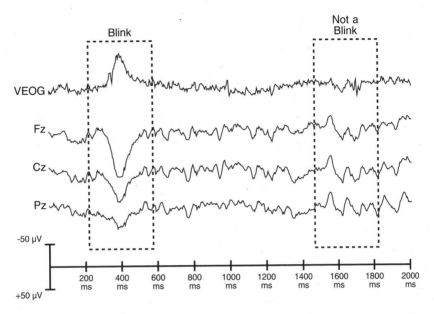

Figure 4.8   Recordings from a vertical EOG (VEOG) electrode located under the left eye and EEG electrodes located at Fz, Cz, and Pz, with a right mastoid reference for all recordings. A blink can be seen at approximately 400 ms, and it appears as a negative deflection at the VEOG electrode and as a positive deflection at the scalp electrodes. Note that the deflection is quite large at Fz and then becomes smaller at Cz and even smaller at Pz. The area labeled "Not a Blink" contains moderately large voltage deflections in all of these channels, but these deflections do not reflect a blink because the polarity is not inverted at the VEOG electrode relative to the scalp electrodes. Negative is plotted upward.

at several locations on the scalp (all are referenced to a mastoid electrode). The eyeblink response consists primarily of a monophasic deflection of 50–100 µV with a typical duration of 200–400 ms. Perhaps the most important characteristic of the eyeblink response, however, is that it is opposite in polarity for sites above versus below the eye (compare, for example, the VEOG and Fz recordings in figure 4.8). This makes it possible to distinguish between a blink, which would produce opposite-polarity voltage shifts above versus below the eye, and a true EEG deflection, which would typically produce same-polarity voltage shifts above and below the eye. The right side of figure 4.8 shows an example of a true EEG deflection, where similar-polarity deflections appear at the VEOG and Fz sites.

Because of the polarity reversal exhibited by blinks, you should always be suspicious of an experimental effect that is opposite in polarity at electrode sites above versus below the eyes. Although such a pattern is possible for a true ERP effect, it should serve as a warning signal indicating that the averaged ERP waveforms may be more contaminated by blinks in one condition than in the other.

Reducing the occurrence of an artifact is always better than rejecting trials with artifacts, and there are several ways to reduce the number of blinks. The first is to ask subjects who normally wear contact lenses—which cause a great deal of blinking—to wear their glasses instead of their contact lenses. These individuals tend to blink more than average even when wearing glasses, and it is therefore useful to keep a supply of eyedrops handy (although you should offer them only to individuals who normally use eyedrops, and you should use single-use bottles to avoid infection risks). Another method for reducing the occurrence of blinks is to use short trial blocks of 1–2 minutes, thus providing the subjects with frequent rest breaks for blinking (this also helps to keep the subjects more alert and focused on the task). The use of such short trial blocks tends to slow down the progress of a recording session, but I have found that this slowing can be mitigated by using trial blocks of 5–6 minutes that are subdivided into "miniblocks" of 1–2 minutes, with automatic breaks of 20–30 seconds interposed between the miniblocks and somewhat longer, experimenter-controlled breaks between the full blocks.

If you see a lot of blinks (or eye movements), it is important to let the subject know. Don't be shy about telling subjects that they need to do a better job of controlling these artifacts. My students tell me that it took them a long time to become comfortable doing this, but you really need to do it, even if it makes you uncomfortable at first.

Blinks are relatively easy to detect on single trials, and a peak-to-peak amplitude measure is usually an adequate artifact rejection function (a simple voltage threshold, however, is clearly inadequate, because a threshold that is sufficiently low to reject all blinks often leads to a large number of false alarms). The peak-to-peak amplitude function can sometimes be "fooled" by slow voltage shifts that cause one end of the epoch to be substantially different in voltage from the other end, and high-frequency noise (e.g., muscle activity) can exacerbate this. Both of these problems can be minimized by a measure that I call a "step" function, which basically looks for step-like changes in voltage. This function is similar to performing a cross-correlation between the EEG/EOG epoch and a function that looks like a step (i.e., a flat low interval followed by a flat high interval). One first defines the width of the step, with a typical value of 100 ms. For each point in the epoch, the mean value of the preceding 100 ms is then subtracted from the mean value of the subsequent 100 ms (or whatever the desired step width is). After this has been computed for each point in the epoch, the largest value is compared with the threshold to determine whether the trial should be rejected. This computationally simple procedure is effective for two reasons. First, averaging together the voltage of a 100-ms interval essentially filters out any high-frequency activity. Second, computing the difference between successive 100-ms intervals minimizes the effects of any gradual changes in voltage, which corresponds with the fact that a blink produces a relatively sudden voltage change.

Whenever possible, one should obtain recordings from an electrode below the eye and an electrode above the eye, with both electrodes referenced to a common, distant site (e.g., an EOG electrode located below one of the two eyes and a frontal EEG site, both referenced to a mastoid electrode). This makes it possible to take advantage of the inversion of polarity exhibited by blinks for sites above and below the eyes, which is especially useful when inspecting the single-trial data during the assessment of the adequacy of the artifact rejection function and the threshold. As discussed above, this also makes it possible to determine whether the ERP averages are contaminated by blinks, which leads to an inversion in polarity for sites above versus below the eyes. In addition, this montage makes it possible to implement an even more powerful artifact rejection function, which I call the "differential step" function. This function is just like the step function, except that one computes values at each time point for a location below and a location above the eyes and then subtracts these two values from each other. The maximum value of this difference is then compared with a threshold to determine whether a given trial should be rejected. The subtraction process ensures that a large value is obtained only when the voltage is changing in opposite directions for the electrodes above and below the eyes. A simpler but nearly equivalent alternative is to apply the standard step function to a bipolar recording in which the electrode beneath the eye is the active site and the electrode above the eye is the reference.

Eye Movements   Like blinks, eye movements are a result of the intrinsic voltage gradient of the eye, which can be thought of as a dipole with its positive end pointing toward the front of the eye. When the eyes are stationary, this dipole creates a constant DC voltage gradient across the scalp, which the high-pass filter of the amplifier eliminates. When the eyes move, the voltage gradient across the scalp changes, becoming more positive at sites that the eyes have moved toward. For example, a leftward eye movement causes a positive-going voltage deflection on the left side of the scalp and a negative-going voltage on the right side of the scalp. It is easiest to observe these deflections with bipolar recordings, in which an electrode lateral to one eye is the active site and an electrode lateral to the other eye is the reference site.

Hillyard and Galambos (1970) and Lins et al. (1993a) have systematically measured the average size of the saccade-produced deflection. These studies yielded the following findings: (a) the voltage deflection at a given electrode site is a linear function of the size of the eye movement, at least over a 15-degree range of eye movements; (b) a bipolar recording of the voltage between electrodes at locations immediately adjacent to the two eyes will yield a deflection of approximately 16 µV for each degree of eye movement; and (c) the voltage falls off in a predictable manner as the distance between the electrode site and the eyes increases (see tables V and VI of Lins et al., 1993, for a list of the propagation factors for a variety of standard electrode sites).

Note also that eye movements cause the visual input to shift across the retina, which creates a visual ERP response (saccadic suppression mechanisms make us unaware of this motion, but it does create substantial activity within the visual system). This saccade-induced ERP depends on the nature of the stimuli that are visible when the eyes move, just as the ERP elicited by a moving stimulus varies as a function of the nature of the stimulus. Procedures that attempt to correct for the EOG voltages produced by eye movements—discussed at the end of this chapter—cannot correct for these saccade-induced ERP responses.

Unless the subject is viewing moving objects or exhibiting gradual head movements, the vast majority of eye movements will be *saccades*, sudden ballistic shifts in eye position. The top three waveforms of figure 4.9 show examples of eye movement recordings. In the absence of noise, a saccade would consist of a sudden transition from the zero voltage level to a nonzero voltage level, followed by a gradual return toward zero caused by the amplifier's high-pass filter (unless using DC recordings). In most cases, subjects make a saccade in one direction and then another to return to the fixation point, which would lead to a boxcar-shaped function in a DC recording and a sloped boxcar-shaped function when using a high-pass filter. This characteristic shape can be used to distinguish small eye movements from normal EEG deflections when visually inspecting the individual trials.

Because of the approximately linear relationship between the size of an eye movement and the magnitude of the corresponding EOG deflection, large eye movements are relatively easy to detect on single trials, but small eye movements are difficult to detect. If one uses a simple voltage threshold to detect and reject eye movement artifacts, with a typical threshold of 100 µV, eye movements as large as 10 degrees can escape detection (e.g., if the voltage

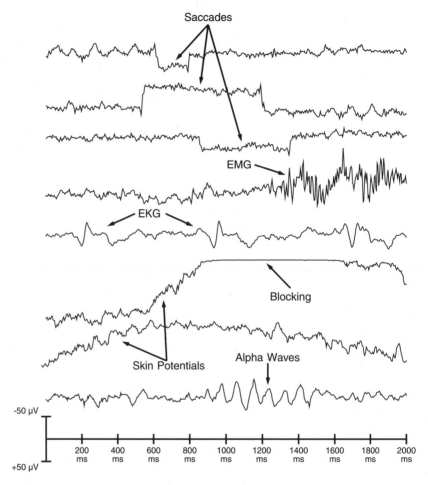

Figure 4.9   EOG and EEG recordings showing several types of artifacts. The saccades were recorded from a horizontal EOG configuration, with the active electrode adjacent to the right eye and the reference electrode adjacent to the left eye. The EMG, blocking, and skin potential artifacts were recorded at Cz with a right mastoid reference. The EKG artifacts were recorded at the left mastoid with a right mastoid reference. The alpha waves were recorded at O2 with a right mastoid reference. Negative is plotted upward.

starts at -80 µV, a 10-degree eye movement in the appropriate direction will cause a transition to $+80$ µV, which would be entirely within the allowable window of ±100 µV). Of course, a 10-degree eye movement greatly changes the position of the stimulus on the retina, which can be an important confound, and the resulting voltage deflection is quite large relative to the size of a typical ERP component, even at scalp sites fairly far from the eyes. However, using a lower threshold will lead to a large number of false alarms, and a simple threshold function is therefore an inadequate means of rejecting trials with eye movements. Peak-to-peak amplitude is somewhat superior to a threshold, but can be fooled by slow shifts in voltage. The step function described above for detecting blinks is better yet, because it is sensitive to temporally circumscribed shifts in voltage. Perhaps the best approach, however, would be to adapt the algorithms developed by vision researchers in models of edge detection, which is a conceptually similar problem. To my knowledge, however, no one has yet applied these algorithms to the problem of saccade detection.

Using a step function, it is possible to detect eye movements as small as 1 to 2 degrees on individual trials, but the S/N ratio of the EOG signal makes it impossible to detect smaller eye movements without an unacceptably large number of false alarms. However, it is sometimes possible to use averaged EOG waveforms to demonstrate that a given set of ERPs are uncontaminated by very small systematic eye movements. Specifically, if different trial types would be expected to elicit eye movements in different directions, you can obtain virtually unlimited resolution by averaging together multiple trials on which the eye movements would be expected to be similar. For example, if an experiment contains some targets in the LVF and other targets in the RVF, one can compute separate averaged EOG waveforms for the LVF and RVF targets and compare these waveforms. Any consistent differential eye movements will lead to differences in the averaged EOG waveforms, and even very small eye movements can be observed due to the improvement in S/N ratio produced by the averaging process.

This procedure will not allow individual trials to be rejected, nor will it be useful for detecting eye movements that are infrequent or in the same direction for both LVF and RVF targets. However, it can be useful when combined with the rejection of individual trials with large eye movements in a two-tiered procedure. The first tier consists of the rejection of individual trials with large saccades (> 1 degree) by means of the step function. You can then examine residual EOG activity in the averaged EOG waveforms, and exclude any subjects with differential EOG activity exceeding some criterion (e.g., 1.6 µV, corresponding to 0.1 degree) from the final data set.

Note that the techniques described above are useful for detecting saccades, but are not usually appropriate for detecting slow shifts in eye position or for assessing absolute eye position. To assess these, it is usually necessary to record the EOG using a DC amplifier, although you can use high-pass filtered EOG recordings for this purpose under some circumstances (Joyce et al., 2002).

Slow Voltage Shifts   Slow voltage shifts are usually caused by a change in the impedance of the skin or the impedance of the electrodes (see *skin potentials* in figure 4.9). There is a small voltage between the superficial and deep layers of skin, and this voltage changes as the impedance changes in accordance with Ohm's Law, which states that voltage is proportional to the product of current and resistance (E = IR; see the appendix). If you increase the resistance of an electrical current without changing the current flow, the voltage must necessarily increase; thus, increasing the resistance actually increases the voltage. Impedance is simply the AC analog of resistance, so increases in impedance also lead to increases in resistance. When subjects sweat (even slightly), this causes a decrease in impedance, and the resulting slow voltage shifts are called skin potentials. The best way to reduce skin potentials is to reduce the impedance of the skin before applying the electrodes. Because electricity preferentially follows the path of least resistance, a change in impedance at one spot on the skin won't influence the overall impedance much if there is also a nearby spot with very low impedance. In general, the greater the initial impedance, the greater will be the changes in impedance due to sweating (see Picton & Hillyard, 1972). It is also helpful to maintain the recording chamber at a cool temperature and low humidity level. On hot and humid summer days, my lab runs a fan in the recording chamber between trial blocks.

Voltage shifts can also be caused by slight changes in electrode position, which are usually the result of movements by the subject. A change in electrode position will often lead to a change in impedance, thus causing a sustained shift in voltage. You can reduce this type of artifact by making sure that the subject is comfortable and does not move very much (a chin rest is helpful for this). If electrodes are placed at occipital sites, the subject should not place the back of his or her head against the back of the chair. You can also greatly reduce slow voltage shifts by using a high-pass filter during data acquisition, which will cause the voltage to return gradually toward 0 μV whenever a shift in voltage occurs (see chapter 5 for details).

It is not usually necessary to reject trials with slow voltage shifts, as long as they are rare. If the voltage shifts are slow and random, they shouldn't distort the averaged ERPs very much. However, a movement in the electrodes will sometimes cause the voltage to change suddenly to a new level, and you can detect this by means of a peak-to-peak amplitude function or a step function applied to each of the EEG channels (you'll want to keep the threshold fairly high to avoid rejecting trials with large ERP deflections, however). In some cases (e.g., when using very long epochs), you may wish to reject trials with gradual shifts in voltage, and you can accomplish this by computing the slope of the EEG across the trial and rejecting trials with slopes above some threshold.

Amplifier Saturation   Slow voltage shifts may sometimes cause the amplifier or ADC to saturate, which causes the EEG to be flat for some period of time (this is also called *blocking*). If this happens frequently, you should simply use a lower gain on the amplifier; if it never happens, you may wish to use a higher gain. As figure 4.9 illustrates, amplifier blocking is relatively easy to spot visually, because the EEG literally becomes a flat line. You could reject trials with amplifier saturation by finding trials in which the voltage exceeds some value that is just below the amplifier's saturation point, but in practice this would be difficult, because the saturation point may vary from channel to channel and may even vary over time. Another possibility would be to determine if there are a large number of points with identical voltages within each trial, but this isn't quite optimal because the voltages might not be exactly the same from moment to moment. A better procedure is to use a function that I call the *X-within-Y-of-peak* function, which Jon Hansen developed at UCSD. This function first finds the maximum EEG value within a trial (the peak), and then counts the number of points that are at or near that maximum. X is the number of points, and Y defines how close a value must be to the peak to be counted. For example, you might want to reject any trial in which thirty or more points are within 0.1 μV of the peak (i.e., $X = 30$ and $Y = 0.1$ μV). Of course, you must apply the same function to both the positive peak voltage and the negative peak voltage, and you should apply it to every channel.

Alpha Waves   Alpha waves are oscillatory EEG deflections around 10 Hz that are largest at posterior electrode sites and occur most frequently when subjects are tired (see figure 4.9). The best way to reduce alpha waves is to use well-rested subjects, but some individuals have substantial alpha waves even when they are fully alert. Alpha waves can be particularly problematic when using a constant stimulus rate, because the alpha rhythm can become entrained to the stimulation rate such that the alpha waves are not reduced by the averaging process. Thus, it is useful to include a jitter of at least ±50 ms in the intertrial interval.

It is not usually worthwhile to reject trials with alpha waves: because ERPs can contain voltage deflections in the 10-Hz range, it is possible that trials with large ERPs will be rejected along with trials containing alpha artifacts. If it is necessary to reject trials with alpha, the best procedure is to compute the amplitude at 10 Hz on each trial and reject trials on which the amplitude exceeds some threshold value.

Muscle and Heart Activity   The voltages created during the contraction of a muscle are called the electromyogram or EMG (see figure 4.9). These voltages are very high in frequency, and much of the EMG is usually eliminated by the amplifier's low-pass filter. You can also minimize the EMG by asking the subject to relax the muscles of the neck, jaw, and forehead and by providing a chinrest or some other device that reduces the load on the neck muscles.[4] As discussed above, it is also possible for the subject to recline in a comfortable chair, but this can cause movements of the posterior electrodes, resulting in large artifactual voltage shifts. Relaxation of the muscles below the neck is usually not important, because the EMG tends not to propagate very far.

It is not usually necessary to reject trials with EMG, assuming that you have taken appropriate precautions to minimize the EMG. However, if it is necessary to reject trials with EMG activity, you can detect EMG in several ways. The best method is to perform a Fourier transform on each trial and calculate the amount of high-frequency power (e.g., power above 100 Hz). A simpler method is to calculate the difference in voltage between every consecutive pair of points in a given trial and reject the trial if the largest of these differences exceeds a particular value.

Note that some stimuli will elicit reflexive muscle twitches. These are particularly problematic because they are time-locked to the stimulus and are therefore not attenuated by the averaging process. These also tend to be sudden, high-frequency voltage changes, but they are usually limited to a very short time period and are therefore difficult to detect by examining the high-frequency power across the entire trial. To reject these artifacts, it is best to look for sudden shifts in voltage during the brief time period during which they are likely to occur (usually within 100 ms of stimulus onset).

The beating of the heart (the EKG) can also be observed in EEG recordings in some subjects; figure 4.9 shows its distinctive shape. The EKG is usually picked up by mastoid electrodes, and if a mastoid is used as a reference, the EKG is seen in inverted form in all of the electrode sites. The EKG can sometimes be reduced by slightly shifting the position of the mastoid electrode, but usually there is nothing that can be done about it. In addition, this artifact usually occurs approximately once per second during the entire recording session, so rejecting trials with EKG deflections will usually lead to the rejection of an unacceptably large proportion of trials. Fortunately, this artifact is almost never systematic, and it will simply decrease the overall S/N ratio. In other words, there isn't much you can do, and it's not usually a significant problem, so don't worry about it.

The previous paragraph raises an important point. If you see an artifact or some type of noise equally in all of your EEG channels, it is probably being picked up by the reference electrode. Most artifacts and noise sources will be more prominent at some electrodes than at others, but any signals picked up by the reference electrode will appear in inverted form in all electrodes that use that reference. However, if you are using bipolar recordings for some of your channels (e.g., for EOG recordings), these recordings will not have artifacts or noise arising from the main reference electrode. This can help you identify and eliminate the sources of noise and artifacts.

Artifact Correction

Artifact rejection is a relatively crude process, because it completely eliminates a subset of trials from the ERP averages. As Gratton, Coles, and Donchin (1983) discussed, there are three potential problems associated with rejecting trials with ocular artifacts. First, in some cases, discarding trials with eye blinks and eye movements might lead to an unrepresentative sample of trials. Second, there are some groups of subjects (e.g., children and psychiatric patients) who cannot easily control their blinking and eye movements, making it difficult to obtain a sufficient number of artifact-free trials. Third, there are some experimental paradigms in which blinks and eye movements are integral to the tasks, and rejecting trials with these artifacts would be counterproductive. Under these conditions, it would be useful to be able to subtract away the voltages due to eye blinks and eye movements rather than rejecting trials with these artifacts.

Researchers have developed several artifact correction procedures for this purpose (e.g., Berg & Scherg, 1991a, 1994; Gratton, Coles, & Donchin, 1983; Lins et al., 1993b; Verleger, Gasser, & Moecks, 1982). When the eyes blink or move, voltages are created around the eyes that propagate to the scalp electrodes, and the voltage recorded at a given site will be equal to the value at the eyes multiplied by a propagation factor, plus any EEG activity present at that site. The simplest way to correct for eye artifacts is to calculate the propagation factor between the eyes and each of the scalp electrodes and subtract a corresponding proportion of the recorded EOG activity from the ERP waveform at each scalp site. For example, Lins and colleagues (1993a) found that 47 percent of the voltage present in an EOG recording propagated to the Fpz electrode, 18 percent to the Fz electrode, and 8 percent to the Cz electrode. To subtract away the EOG contribution to the averaged ERP waveforms at these electrode sites, it would be possible to subtract 47 percent of the EOG waveform from the Fpz electrode, 18 percent from the Fz electrode, and 8 percent from the Cz electrode.

There is a very significant problem with this approach, however. Specifically, the EOG recording contains brain activity in addition to true ocular activity and, as a result, the subtraction procedure ends up subtracting away part of the brain's response as well as the ocular artifacts. There are additional problems with this simple-minded subtraction, such as the assumption that the propagation factors will be the same for eye blinks and eye movements (a problem first addressed by Gratton, Coles, & Donchin, 1983). More sophisticated versions of this approach address these additional problems, and can work fairly effectively. For example, one can use dipole modeling procedures to isolate the ocular activity (Berg & Scherg, 1991b), which works fairly well because the approximate locations of the dipoles are known in advance.

Although these artifact correction techniques can be useful or even indispensable for certain tasks and certain types of subjects, they have some significant drawbacks. First, some of these techniques can significantly distort the ERP waveforms and scalp distributions, making the data difficult to interpret. On the basis of a detailed comparison of several techniques, Lins et al. (1993b) concluded that source analysis procedures provided the least distortion, and other techniques (such as those of Gratton, Coles, & Donchin, 1983; Verleger, Gasser, & Moecks, 1982) can yield significant distortion. However, even the source analysis procedures may yield some distortion, especially when non-optimal parameters are used.

A newer and promising approach is to use independent components analysis (ICA). This approach is well justified mathematically, and recent studies demonstrated that this technique works very well at removing blinks, eye movements, and even electrical noise (Jung, Makeig, Humphries et al., 2000; Jung, Makeig, Westerfield et al., 2000) (see also the similar technique developed by Joyce, Gorodnitsky, & Kutas, 2004). However, these studies were conducted by the group who originally developed ICA, so they may not have been motivated to find conditions under which ICA performs poorly. In particular, this approach assumes that the time course of the artifacts is independent of the time course of the ERP activity, which may not always (or even usually) be a correct assumption. For example, if detecting a target leads to both a P3 wave and a blink, the blink and the P3 wave will have correlated time courses, and this could lead to inaccurate artifact correction. Until an independent laboratory rigorously tests this technique, it will be difficult to know whether this sort of situation leads to significant distortions.

A second problem with artifact correct techniques is that these techniques may require significant additional effort. For example,

Lins et al. (1993b) recommended that recordings should be obtained from at least seven electrodes near the eyes. In addition, one must conduct a set of calibration runs for each subject and carry out extensive signal processing on the data. Thus, it is important to weigh the time saved by using artifact correction procedures against the time required to satisfactorily implement these procedures.

A third problem with these techniques is that they cannot account for the changes in sensory input caused by blinks and eye movements. For example, if a subject blinks at the time of a visual stimulus, then this stimulus may not be seen properly, and this obviously cannot be accounted for by artifact correction techniques. In addition, as the eyes move, the visual world slides across the retina, generating a sensory ERP response. Similarly, eye blinks and eye movements are accompanied by motor ERPs. Artifact correction procedures do not typically address these factors, which are especially problematic when task-relevant stimuli trigger the blinks or eye movements.

Because of these limitations, I would recommend against using artifact correction procedures unless the nature of the experiment or subjects makes artifact rejection impossible. When artifact correction is necessary, I would recommend using one of the newer and less error-prone techniques, such as ICA or the source localization techniques Lins et al. (1993b) discuss. Moreover, I would strongly recommend against using the simpler techniques that are often available in commercial ERP analysis packages (such as the procedure of Gratton et al., 1983). When you use these techniques, it is difficult to know the extent to which the artifact correction procedures distort the results.

Suggestions for Further Reading

The following is a list of journal articles and book chapters that provide useful information about averaging, artifact rejection, and artifact correction.

Berg, P., & Scherg, M. (1994). A multiple source approach to the correction of eye artifacts. *Electroencephalography & Clinical Neurophysiology, 90*(3), 229–241.

Gratton, G., Coles, M. G. H., & Donchin, E. (1983). A new method for off-line removal of ocular artifact. *Electroencephalography and Clinical Neurophysiology, 55,* 468–484.

Hillyard, S. A., & Galambos, R. (1970). Eye movement artifact in the CNV. *Electroencephalography and Clinical Neurophysiology, 28,* 173–182.

Jung, T. P., Makeig, S., Humphries, C., Lee, T. W., McKeown, M. J., Iragui, V., & Sejnowski, T. J. (2000). Removing electroencephalographic artifacts by blind source separation. *Psychophysiology, 37,* 163–178.

Jung, T. P., Makeig, S., Westerfield, M., Townsend, J., Courchesne, E., & Sejnowski, T. J. (2000). Removal of eye activity artifacts from visual event-related potentials in normal and clinical subjects. *Clinical Neurophysiology, 111,* 1745–1758.

Lins, O. G., Picton, T. W., Berg, P., & Scherg, M. (1993). Ocular artifacts in EEG and event-related potentials I: Scalp topography. *Brain Topography, 6,* 51–63.

Lins, O. G., Picton, T. W., Berg, P., & Scherg, M. (1993). Ocular artifacts in recording EEGs and event-related potentials. II: Source dipoles and source components. *Brain Topography, 6,* 65–78.

Picton, T. W., Linden, R. D., Hamel, G., & Maru, J. T. (1983). Aspects of averaging. *Seminars in Hearing, 4,* 327–341.

Verleger, R., Gasser, T., & Moecks, J. (1982). Correction of EOG artifacts in event-related potentials of the EEG: Aspects of reliability and validity. *Psychophysiology, 19*(4), 472–480.

Woldorff, M. (1988). Adjacent response overlap during the ERP averaging process and a technique (Adjar) for its estimation and removal. *Psychophysiology, 25,* 490.

5 Filtering

This chapter discusses the application of filters to the EEG during data acquisition and to ERP waveforms before and after the averaging process. It is absolutely necessary to use filters during data acquisition, and it is very useful to apply filters offline as well, but filtering can severely distort ERPs in ways that ERP researchers frequently do not appreciate. For example, filters may change the onset and duration of an ERP component, may make monophasic waveforms appear multiphasic, may induce artificial oscillations, and may interfere with the localization of generator sources. This chapter will explain how these distortions arise and how they can be prevented. To avoid complex mathematics, I will simplify the treatment of filtering somewhat in this chapter, but there are several books on filtering that the mathematically inclined reader may wish to read (e.g., Glaser & Ruchkin, 1976). Note also that the term *filter* can refer to any of a large number of data manipulations, but this chapter will be limited to discussing the class of filters that ERP researchers typically use to attenuate specific ranges of frequencies, which are known as *finite impulse response filters*.

ERP waveforms are generally conceptualized and plotted as *time-domain* waveforms, with time on the X axis and amplitude on the Y axis. In contrast, filters are typically described in the *frequency-domain*, with frequency on the X axis and amplitude or power on the Y axis.[1] Because ERP researchers are typically more interested in temporal information rather than frequency information and because temporal information may be seriously distorted by filtering, it is important to understand filtering as a time-domain operation as well as a frequency-domain operation.[2] This chapter therefore describes how filters operate in both the time and frequency domains, as well as the relationship between these domains. I will focus primarily on the *digital* filters that are used for offline filtering, but the principles discussed here also apply to the *analog* filters that are found in amplifiers. My goal in this chapter is to provide an intuitive understanding of how filters work, and I have therefore minimized the use of equations as much as possible and limited the equations to simple algebra. If you understand how summations work (i.e., the Σ symbol), then you know enough math to understand this chapter.

Even though an advanced math background is not necessary to understand this chapter, some of the concepts are pretty complicated. As a result, you may not want to spend the time required to understand exactly how filters work and exactly why they distort ERP waveforms in particular ways, especially on your first pass through the book. If so, you can just read the next two sections ("Why Are Filters Necessary" and "What Everyone Should Know About Filtering"). These sections will provide enough information to keep you from getting into too much trouble with filters, and you can read the rest of the chapter later.

Why Are Filters Necessary?

The most important message of this chapter is that filters can substantially distort ERP data. Given that this is true, it is useful to ask why filters are necessary in the first place. There are two answers to this question, the first of which is related to the *Nyquist Theorem*, discussed previously in chapter 3. This theorem states that it is possible to convert a continuous analog signal (such as the EEG) into a set of discrete samples without losing any information, as long as the rate of digitization is at least twice as high as the highest frequency in the signal being digitized. This is fundamentally important for the data acquisition process, because it means that we can legitimately store the EEG signal as a set of discrete samples on a computer. However, this theorem also states that if the original signal contains frequencies that are more than twice as high as the digitization rate, these very high frequencies will appear as artifactual *low* frequencies in the digitized data (this is called *aliasing*). Consequently, EEG amplifiers have filters that one can use to suppress high frequencies; these filters are generally set to attenuate frequencies that are higher than a half of the sampling rate. For example, in a typical cognitive ERP experiment, the digitization rate might be 250 Hz, and it would therefore be necessary to make sure that everything above 125 Hz is filtered. It would be tempting to choose a filter cutoff frequency of 125 Hz, but a cutoff frequency of 125 Hz just means that the power (or amplitude) has been reduced by 50 percent at 125 Hz, and there will be considerable remaining activity above 125 Hz. It is therefore necessary to select a substantially lower value such as 80 Hz (in practice, the digitization rate should be at least three times as high as the cutoff value of the filter).

The second main goal of filtering is the reduction of noise, and this is considerably more complicated. The basic idea is that the EEG consists of a signal plus some noise, and some of the noise is sufficiently different in frequency content from the signal that it can be suppressed simply by attenuating certain frequencies. For example, most of the relevant portion of the ERP waveform in a typical cognitive neuroscience experiment consists of frequencies between 0.01 Hz and 30 Hz, and contraction of the muscles leads to an EMG artifact that primarily consists of frequencies above 100 Hz; consequently, the EMG activity can be eliminated by suppressing frequencies above 100 Hz and this will cause very little change to the ERP waveform. However, as the frequency content of the signal and the noise become more and more similar, it becomes more and more difficult to suppress the noise without significantly distorting the signal. For example, alpha waves can provide a significant source of noise, but because they are around 10 Hz, it is difficult to filter them without significantly distorting the ERP waveform. Moreover, even when the frequency content of the noise is very different from that of the signal, the inappropriate application of a filter can still create significant distortions.

In addition to suppressing high frequencies, filters are also used in most experiments to attenuate very low frequencies. The most common use of such filters is to remove very slow voltage changes of non-neural origin during the data acquisition process. Specifically, factors such as sweating (which creates skin potentials) and drifts in electrode impedance can lead to slow, sustained changes in the baseline voltage of the EEG signal, and it is usually a good idea to remove these slow voltage shifts by filtering frequencies lower than approximately 0.01 Hz. This is especially important when obtaining recordings from patients or from children, because head and body movements are one common cause of these sustained shifts in voltage. If these shifts are not eliminated, they may cause the amplifier to saturate and data to be lost. Even if they do not cause amplifier saturation, they are very large voltages and may cause large distortions in the average ERP waveforms. Thus, it is almost always a good idea to filter the very low frequencies (< 0.01 Hz).

What Everyone Should Know about Filtering

Any waveform can be decomposed into a set of sine waves of various frequencies and phases, and the waveform can be reconstructed by simply summing these sine waves together. Filters are usually described in terms of their ability to suppress or pass various different frequencies. The most common types of filters are: (1) *low-pass filters*, which attenuate high frequencies and pass low frequencies; (2) *high-pass filters*, which attenuate low frequencies and pass high frequencies; (3) *bandpass filters*, which attenuate both high and low frequencies, passing only an intermediate range of frequencies; and (4) *notch filters*, which attenuate some narrow band of frequencies and pass everything else.

Filters and Amplitude Attenuation

The properties of a filter are usually expressed by its *transfer function*, the function that determines how the input of the filter is "transferred" to the output of filter. The transfer function can be broken down into two components, a frequency response function that specifies how the filter changes the amplitude of each frequency, and a phase response function that specifies how the filter changes the phase of each frequency. Figure 5.1A shows an example of the frequency response function of a low-pass filter with a half-amplitude cutoff at 30 Hz, and it also shows how this filter suppresses 60-Hz noise in an actual ERP waveform. In ERP research, a filter is often described only in terms of its *half-amplitude cutoff*, which is the frequency at which the amplitude is cut by 50 percent (sometimes the cutoff frequency is specified in terms of power rather than amplitude; amplitude reaches 71 percent when power reaches 50 percent). As figure 5.1A illustrates, however, there is quite a bit of attenuation at frequencies below the half-amplitude cutoff, and the attenuation is far from complete for frequencies quite a bit higher than the half-amplitude cutoff. Nonetheless, this filter effectively suppresses the 60-Hz noise in the waveform while retaining the waveform's basic shape.

Figure 5.1B shows the frequency response function of a high-pass filter with a half-amplitude cutoff at approximately 2.5 Hz and the effect of this filter on an ERP waveform. As you can see, the large, broad upward deflection (similar to a P3 wave) is greatly reduced in amplitude by this filter, whereas the initial, higher frequency deflections (similar to P1 and N1 waves) are not reduced as much. However, the amplitudes of the earlier components are influenced somewhat, and small artifactual peaks have been created at the beginning and end of the waveform. This leads to the most important point of this section: *filters can significantly distort ERP waveforms, changing the amplitude and timing of the ERP components and adding artifactual peaks.*

High-pass filters are often described in terms of their *time constants* rather than their half-amplitude cutoffs. As figure 5.1C shows, if the input to a high-pass filter is a constant voltage, the output of the filter will start at this voltage and then gradually return toward zero, and the filter's time constant is a measure of the rate at which this occurs. The decline in output voltage over

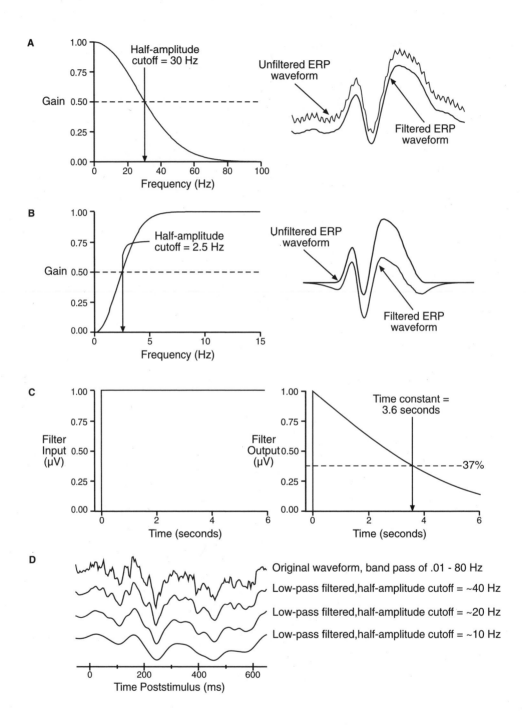

time is exponential, so the value never quite reaches zero. Consequently, the time constant is expressed as the time required for the filter's output to reach $1/e$ (37 percent) of the starting value. As the half-amplitude cutoff becomes higher, the time constant becomes shorter. If you know the half-power cutoff frequency of a high-pass filter (f_c, the frequency at which the filter's output is reduced by 3 dB), the time constant can be computed as $1/(2\pi f_c)$.

Filters and Latency Shift

So far, we have focused on the effects of filters on the amplitude at each frequency, but it also important to keep in mind that filters usually influence the phase (time shift) at each frequency as well. In particular, all analog filters (such as those in EEG amplifiers) shift the latency of the signal, and most analog filters shift the latency by different amounts for different frequencies. These shifts become more pronounced as the cutoff frequency of a low-pass filter is made lower and the cutoff frequency of a high-pass filter is made higher. In contrast, most digital filters (filters that are applied offline) do not cause a phase shift. Thus, it is usually best to do as little filtering as possible prior to digitizing the data and to do most of the filtering offline.

How Filters Are Used

Some filtering is essential in amplifying and digitizing the EEG signal. First, it is necessary to filter out high frequencies with a low-pass filter before digitizing the data to ensure that the

◀ **Figure 5.1** Basics of filtering. (A) Frequency response function of a low-pass filter with a half-amplitude cutoff at 30 Hz, and application of this filter to an ERP waveform. (B) Frequency response function of a high-pass filter with a half-amplitude cutoff at 2.5 Hz, and application of this filter to an ERP waveform. (C) Example of the time constant of a high-pass filter. A constant input to the filter is shown on the left, and the decaying output of the filter is shown on the right. (D) Noisy ERP waveform low-pass filtered with various half-amplitude cutoffs.

sampling rate is at least twice as high as the highest frequency in the signal. Second, it is almost always necessary to filter out very low frequencies (e.g., <0.01 Hz) so that slow, non-neural electrical potentials (e.g., skin potentials) do not bring the signal outside of the operating range of the amplifier and analog-to-digital converter. ERP researchers also frequently use a *line frequency filter*, a notch filter that filters out noise generated by AC electrical devices (usually 50 Hz or 60 Hz). As I will discuss later in this chapter, however, notch filters significantly distort the data and should be avoided if possible. Chapter 3 described some strategies for eliminating electrical noise at the source, but if it is impossible to reduce line-frequency noise sufficiently, a low-pass filter with a half-amplitude cutoff at 30 Hz can effectively filter out the noise without distorting the data as much as a notch filter.

ERP researchers sometimes use low-pass and high-pass filters to "clean up" noisy data. For example, figure 5.1D shows an ERP waveform that was mildly filtered (.01–80 Hz) prior to amplification and the same waveform after it was low-pass filtered offline with progressively lower cutoff frequencies. As the cutoff frequency decreases, the waveform becomes smoother and generally nicer looking. From this example, you might think that filtering is a good thing. However, the most important thing that I would like to communicate in this chapter is that filtering always distorts the ERP waveform, and the more heavily the data are filtered, the worse the distortion will be.

How Filters Can Distort Your Data

The distortions caused by filtering can be summarized by a key principle that you should commit to memory now and recall every time the topic of filtering comes up: *precision in the time domain is inversely related to precision in the frequency domain.* In other words, the more tightly you constrain the frequencies in an ERP waveform (i.e., by filtering out a broad range of frequencies), the more the ERP waveform will become spread out in time. Figure

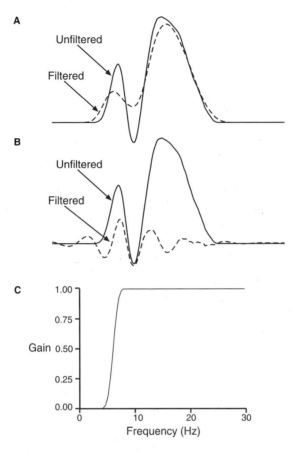

Figure 5.2   Examples of distortions caused by filtering. (A) Effects of low-pass filtering on onset and offset times of an ERP waveform. (B) Artifactual oscillations caused by a high-pass filter. (C) Frequency response function of the filter used in (B).

5.2A illustrates one type of spreading, showing how low-pass filtering an ERP waveform causes the filtered waveform to start earlier and end later than the unfiltered waveform. Low-pass filters almost always have this effect of "smearing out" the waveform and distorting the onset and offset times of the ERP components and experimental effects. Thus, if you see that an experimental effect starts at 120 ms in low-pass filtered waveforms, the effect may have actually started at 150 ms.

As figure 5.2B illustrates, high-pass filters also cause the ERP waveform to become spread out in time, but the distortion in this case consists of a series of up and down sinusoidal deflections. Thus, not only did this particular filter cause artifactual activity to begin well before the actual onset of activity, it created the appearance of oscillations in the waveform. This sort of effect could also cause an experimental effect in a component of one polarity to appear as an effect in an earlier component of opposite polarity (e.g., an increase in N1 amplitude in the original waveform might appear as an increase in P1 amplitude in the filtered waveform). As you can imagine, using this sort of filter might cause someone to completely misinterpret the results of an ERP experiment.

High-pass filters do not always create artifactual oscillations. For example, the high-pass filter shown in figure 5.1B produces a single opposite-polarity artifactual deflection at each end of the ERP waveform rather than a multi-peaked oscillation. The long-lasting and oscillating pattern of distortion the filter shown in figure 5.2B creates is a consequence of the filter's frequency response function, which is shown in figure 5.1C. Whereas the filter in figure 5.1B has a relatively gradual transition between suppressing the lowest frequencies and passing higher frequencies, the filter in figure 5.2B completely blocks a fairly broad set of low frequencies and then suddenly starts passing higher frequencies. That is, the filter in figure 5.2B has more precision in terms of its frequency properties than the filter in figure 5.1B, and this brings us back to the principle that precision in the time and frequency domains are inversely related. That is, more sudden changes in a filter's frequency response function lead to broader temporal distortions in the ERP waveform. Thus, a filter with a really sharp cutoff in its frequency response function might seem to be ideal, but a sharp cutoff usually means that the filter will cause more severe distortion of the ERP waveform than a filter with a gentler cutoff. Makers of commercial ERP systems often tout the sharp cutoffs of their filters, but such filters are usually a bad idea.

If you want to see how a filter might be distorting your ERP waveforms, the easiest thing to do is to pass some sort of known, artificial waveform through the filter and compare the original and filtered waveforms. For example, many amplifiers contain calibrators that produce a single pulse of a square wave, and you can record this and filter it with various types of filters.

Recommendations

Now that I've discussed why filters are used and how they can distort your data, I will make some specific recommendations. These recommendations might not be appropriate for every experiment, but they should be appropriate for the vast majority of ERP experiments in cognitive neuroscience.

First, keep Hansen's Axiom in mind: *There is no substitute for clean data* (see chapter 3). Some minor filtering is necessary when first collecting the data, and a modest amount of additional filtering can be helpful under some conditions. However, filters cannot help you if your data are noisy because of variability across subjects, variability across trials, a small number of trials in your averages, and so on. Filters may make the data *look* better under these conditions, but this will be an illusion that may lead you to draw incorrect conclusions.

Second, during the amplification and digitization process, you should do as little filtering as possible. It's always possible to filter the data more offline, but you can never really "unfilter" data that have already been filtered. The low-pass filter should be set at between one third and one fourth of the sampling rate (and I would recommend a sampling rate of between 200 and 500 Hz for most experiments). For experiments with highly cooperative subjects (e.g., college students), I would recommend using a high-pass filter of 0.01 Hz. Some amplifiers allow you to do no high-pass filtering at all (these are called DC recordings), but unless you are looking at very slow voltage shifts, this will lead to an unacceptable level of noise due to skin potentials and other slow artifacts.

If you are recording from less cooperative subjects (e.g., patients or children), you may need to use a higher high-pass cutoff, such as 0.05 Hz or even 0.1 Hz. The main problem with such subjects is that they may move around a lot, which causes changes in the baseline electrical potential that will slowly resolve over time. This creates a lot of noise in the data that averaging may not sufficiently attenuate, and it can also lead the amplifiers to saturate. But if you use a higher high-pass cutoff, be aware that this is distorting your data somewhat and reducing the amplitude of the lower-frequency components, such as the P3 and N400 waves. Indeed, Duncan-Johnson and Donchin (1979) showed many years ago that relatively high cutoff frequencies lead to a substantial reduction in apparent P3 amplitude.

During amplification, you should avoid using a notch filter (also called a line-frequency filter) to reduce line-frequency noise. These filters can cause substantial distortion of the ERP waveform. As chapter 3 described, there are various precautions you can take to eliminate line-frequency noise before it enters the EEG, and this is the best approach. However, you may sometimes be faced with such a huge level of line-frequency noise that the incoming EEG is completely obscured, and on such occasions you may need to use a notch filter to eliminate this noise.

My third recommendation is to keep offline filtering to a minimum. If the averaged ERP waveforms are a little fuzzy looking, making it difficult to see the experimental effects, it can be helpful to apply a low-pass filter with a half-amplitude cutoff somewhere between 20 and 40 Hz. This can dramatically improve the appearance of the ERP waveforms when you plot them, and the temporal distortion should be minimal (especially if you use a filter with a relatively gentle cutoff). This sort of filtering will also be helpful if you are using peak amplitude or peak latency measures, but it is unnecessary if you are using mean amplitude or fractional area latency measures (see chapter 6 for more details). In fact, because filters spread out an ERP waveform, measuring the mean amplitude in a particular latency range from filtered waveforms (e.g.,

200–250 ms poststimulus) is equivalent to measuring the mean amplitude from a broader range in the original unfiltered waveforms (e.g., 175–275 ms). My laboratory typically uses a low-pass filter with a half-amplitude cutoff at 30 Hz for plotting the data, but no offline filtering for measuring component amplitudes.

My fourth recommendation is to avoid using high-pass filters altogether (except during data acquisition, as described above). High-pass filters are much more likely than low-pass filters to cause major distortions of your ERP waveforms that might lead you to draw incorrect conclusions about your data. There are occasions when high-pass filters can be useful, such as when dealing with overlapping activity from preceding and subsequent stimuli. However, high-pass filters are sufficiently dangerous that you should use them only if you really understand exactly what they are doing. The remainder of this chapter will help provide you with this understanding.

Filtering as a Frequency-Domain Procedure

The key to understanding filters is to understand the relationship between the *time domain* and the *frequency domain*. A time-domain representation of an ERP is simply a plot of the voltage at each time point, as figure 5.3A illustrates. A frequency-domain representation of an ERP is a plot of the amplitude (and phase) at each frequency, as figure 5.3B illustrates. Time-domain and frequency-domain representations contain exactly the same information, viewed from different perspectives, and it is possible to transform one type of representation into the other via *Fourier analysis*, a technique developed in the 1800s by the mathematician Joseph Fourier.

The basic principle of Fourier analysis is that any time-domain waveform can be exactly represented by the sum of a set of sine waves of different frequencies and phases. In other words, it is possible to create any waveform (even a momentary spike or a square wave) by adding together a set of sine waves of varying frequencies

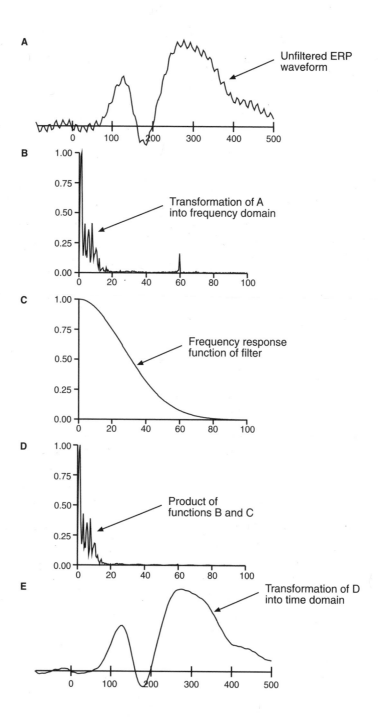

A — Unfiltered ERP waveform

B — Transformation of A into frequency domain

C — Frequency response function of filter

D — Product of functions B and C

E — Transformation of D into time domain

and phases. The *Fourier transform* is a mathematical procedure that takes a time-domain signal (such as an ERP waveform) as an input and computes the amplitudes and phases of the sine waves that would need to be added together to recreate the input waveform. As figure 5.3B illustrates, the output of the Fourier transform is usually shown as a plot of amplitude and phase[3] as a function of frequency. It is also possible to use the *inverse Fourier transform* to convert a frequency-domain representation back into the original time-domain representation, thus obtaining the original voltage × time ERP waveform.

In the context of Fourier analysis, one can conceptualize filtering as a series of three steps:

1. The time-domain ERP waveform is converted into a frequency-domain representation.
2. The to-be-filtered frequency range is set to zero in the frequency-domain representation.
3. The modified frequency-domain representation is converted back into the time domain.

This completely eliminates one set of frequencies from the ERP waveform without influencing the other frequencies. As I will discuss in detail below, a sudden drop-off in the frequency domain leads to some undesirable consequences, so it is necessary to add a slight complication to this three-step procedure. Specifically,

◄ **Figure 5.3** Example of the frequency-domain conceptualization of filtering. (A) Unfiltered ERP waveform, contaminated by substantial noise at 60 Hz. (B) Transformation of (A) into the frequency domain, with a clear peak at 60 Hz. (C) Frequency response function of a filter that can be used to remove the 60-Hz noise from (B) while retaining most of the signal, which is primarily confined to frequencies below 20 Hz. (D) Product of (B) and (C); for each frequency point, the magnitude in (B) is multiplied by the magnitude in (C). Note that (D) is nearly identical to (B) in the low frequency range, but falls to zero at high frequencies. (E) Transformation of (D) back into the time domain, where it closely resembles the original ERP waveform in (A), except for the absence of the 60-Hz noise. Note that the phase portion of the frequency-domain plots has been omitted here for the sake of simplicity, although the phase information is crucial for transforming between the time and frequency domains.

rather than setting some range of frequencies to zero and leaving the other frequencies untouched, it is useful to shift gradually from no attenuation to complete attenuation. The function that defines this gradual shift in attenuation as a function of frequency is called the *frequency response function* of a filter. More specifically, the frequency response function contains a scaling factor for each frequency that represents the extent to which that frequency will be passed by the filter, with a value of 1 for frequencies that will be unaffected by the filter, a value of zero for frequencies that will be completely suppressed, and an intermediate value for frequencies that will be partially attenuated. This function is used in step two of the filtering process: rather than setting some frequencies to zero, the frequency response function is multiplied by the frequency-domain representation of the ERP waveform to attenuate each frequency by a specific amount (i.e., each frequency in the ERP is multiplied by the corresponding scaling factor in the frequency response function).

Figure 5.3 illustrates this approach to filtering. Panel A shows a time-domain representation of an unfiltered ERP waveform, which contains an obvious 60-Hz noise oscillation. Panel B shows the Fourier transform of this waveform. Note that most of the power is at low frequencies (< 20 Hz), with a spike at 60 Hz corresponding to the noise oscillation in the time-domain waveform. Panel C shows the frequency response function of a filter that passes the very lowest frequencies and then gradually falls off at higher frequencies. Panel D shows the result of multiplying the frequency-domain representation of the ERP waveform (i.e., panel B) by the frequency response function of the filter (i.e., panel C). Note that the result is very similar to the waveform in panel B, but the spike at 60 Hz is missing. Panel E shows the result of applying the inverse Fourier transform to the waveform in panel D, which converts the frequency-domain representation back into the time domain. The result is an ERP waveform that is very similar to the original waveform in panel A, but without the 60-Hz noise oscillation.

Note that it is possible to use any set of scaling factors for the frequency response function shown in figure 5.3C. For example, you could set all of the odd-numbered frequencies to 1 and all of the even-numbered frequencies to zero, or you could use the profile of the Rocky Mountains near Banff to determine the scaling factors. In practice, however, frequency response functions are usually smoothly descending (low-pass filters), smoothly ascending (high-pass filters), flat at 1.0 except for a notch (notch filters), or flat at 0.0 for the lowest and highest frequencies with a single peak at intermediate frequencies (band-pass filters).

The above discussion has concentrated on the amplitude of each frequency and has neglected phase. A frequency-domain representation actually has two parts, one representing the amplitude at each frequency and the other representing the phase at each frequency. Filters may shift the phases of the frequencies as well as modulating their amplitudes, and to fully characterize a filter in the frequency domain, it is necessary to specify its transfer function, which specifies both the frequency gain and the phase shift at each frequency. It is usually desirable for the phase portion of the transfer function to be zero for all frequencies, thereby leaving the phases from the ERP waveform unchanged. This is usually possible when digital filters are applied offline, but is impossible when analog filters are applied online (e.g., when low and high frequencies are removed from the EEG during data acquisition). However, certain analog filters have transfer functions in which the phase portion increases linearly as a function of frequency (e.g., Bessel filters), which means that all frequencies are shifted by the same amount of time; the output of such filters can be shifted back by this amount during the averaging process, thus eliminating the phase shift produced by the filter.

A Problem with Frequency-Domain Representations

Although it is mathematically convenient to conceptualize filtering as a frequency-domain procedure, this approach has an important shortcoming when applied to transient ERP waveforms. The difficulty stems from the fact that time-domain waveforms are represented in Fourier analysis as the sum of a set of infinite-duration sine waves, whereas transient ERP waveforms are finite in duration. Although the Fourier representation is accurate in the sense that ERPs can be transformed between the time and frequency domains with no loss of information, it is inaccurate in the sense that ERPs actually consist of finite-duration voltage deflections rather than sets of infinite-duration sine waves. The biologically unrealistic nature of the frequency-domain representation leads to a number of problems that I will discuss in this chapter.

At this point, let's consider a particularly extreme case in which activity at precisely 5 Hz is filtered from an ERP waveform. Completely suppressing the 5-Hz component of an ERP waveform would be equivalent to computing the amplitude and phase in the 5-Hz frequency band and then subtracting a sine wave with this amplitude and phase from the time-domain ERP waveform. After subtraction of this infinite-duration sine wave, the resulting waveform would contain a 5-Hz oscillation during intervals where the unfiltered ERP waveform was flat, such as the prestimulus interval. This is counterintuitive, because filtering out the activity at 5 Hz actually creates a 5-Hz oscillation in the prestimulus interval. Moreover, because the response to a stimulus obviously cannot precede the stimulus in time, this prestimulus oscillation reflects an impossible state of affairs. Thus, because Fourier analysis represents transient ERP waveforms as the sum of infinite-duration sine waves, which is a biologically incorrect representation, the use of analytical techniques based on frequency-domain representations may lead to serious distortions.

As discussed earlier in this chapter, an important rule for understanding the distortion that filters may produce is that there is an inverse relationship between the spread of a signal in the time domain and the spread of the same signal in the frequency domain; a signal that is tightly localized in time will be broadly localized in frequency, and vice versa. For example, a signal with a single sharp spike in the frequency domain will translate into an infinite-duration sine wave in the time domain, and a single sharp spike in the time domain will translate into an even spread across all frequencies in the frequency domain. Because of this inverse relationship between the time and frequency domains, using filters to restrict the range of frequencies in a signal necessarily broadens the temporal extent of the signal. The more narrowly a filter is specified in the frequency domain, the greater the temporal spread. As a result, filtering out a single frequency, as in the 5-Hz filter example described above, leads to an infinite spread of the filtered waveform in time.

Filtering as a Time-Domain Procedure

We will begin our description of time-domain filtering by considering a common-sense approach to suppressing high-frequency noise, such as that present in the ERP waveform shown in figure 5.4A. To attenuate this noise, one could simply average the voltage at each time point with the voltages present at adjacent time points. In other words, the filtered voltage at time point n would be computed as the average of the unfiltered voltages at time points $n - 1$, n, and $n + 1$. Figure 5.4B shows the results of applying such a filter to the ERP waveform in figure 5.4A; this simple filter has clearly eliminated much of the high frequency noise from the waveform. To filter a broader range of frequencies, one can extend this filtering technique by simply averaging together a greater number of points. Figure 5.4C shows the results of averaging seven adjacent points together instead of the three averaged together for figure 5.4B, and this can be seen to further reduce the high frequency content of the waveform. The following simple formula formalizes this method of filtering:

$$fERP_i = \sum_{j=-n}^{n} wERP_{i+j} \tag{5.1}$$

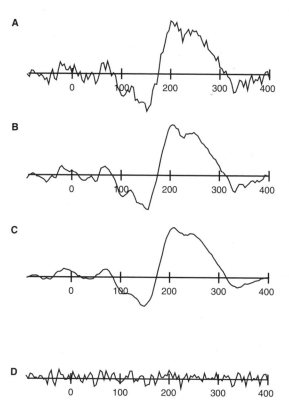

Figure 5.4 Example of filtering an ERP waveform by averaging together the voltages surrounding each time point. (A) Unfiltered ERP waveform, contaminated by substantial high-frequency noise. (B) Result of filtering the waveform in (A) by averaging the voltage at each time point with the voltages at the immediately adjacent time points. (C) Result of filtering the waveform in (A) by averaging the voltage at each time point with the voltages at the three time points on either side. (D) High-pass-filtered waveform, constructed by subtracting the filtered waveform in (C) from the unfiltered waveform (A).

where: $fERP_i$ is the filtered ERP waveform at time i; ERP_i is the unfiltered ERP waveform at time i; n is the number of points to be averaged together on each side of the current time point; and $w = 1/(2n + 1)$ (which is the weighting value).

This equation states that the filtered voltage at a given time point is computed by multiplying each of the n voltages on either side of the current time point and the value at the current time point by the weighting value w and then adding together these weighted values. This is equivalent to averaging these $2n + 1$ points (n points before the current point $+ n$ points after the current point $+$ the current point $= 2n + 1$).

It is worth spending some time to make sure that you understand equation 5.1. Once you understand it, you will have understood the essence of digital filtering.

One can extend this filtering technique in a straightforward manner to attenuate low frequencies instead high frequencies. The unfiltered waveform is equal to the sum of its high frequency components and its low frequency components; as a result, one can filter out the low frequencies by simply subtracting the low-pass-filtered waveform from the unfiltered waveform. Figure 5.4D shows the result of this form of high-pass filtering, which is equal to the waveform in figure 5.4A minus the waveform in figure 5.4C.

When filtering is accomplished by simply averaging together the $2n + 1$ points surrounding each time point, all of the time points being averaged together contribute equally to the filtered value at the current time point, and this reduces the temporal precision of the filtered waveform. To mitigate this problem, one can use a weighted average that emphasizes nearby time points more than distant time points. For example, a three-point filter might use weights of 0.25 for the two adjacent points and a weight of 0.50 for the current point (note that our original filter used equal weights of 0.33 for all three time points). In the general case, we can define an array W that contains $2n + 1$ weights, and recast our filtering formula as:

$$fERP_i = \sum_{j=-n}^{n} W_j ERP_{i+j} \tag{5.2}$$

Our three-point weighting function would then be defined as:

$$W_{-1} = .25, \quad W_0 = .50, \quad W_{+1} = .25$$

For an equal weighting over $2n + 1$ time points, as in our original formulation, the weighting function would be:

$$W_j = \frac{1}{2n + 1}$$

You can use virtually any conceivable weighting function for filtering, and the shape of this function will determine the properties of the filter. As I will describe below, there is a straightforward relationship between the shape of the weighting function and the frequency response function of the filter.

In addition to computing the filtered value at a given time point, it is sometimes useful to consider how the unfiltered value at a given time point influences the filtered values at surrounding time points. If we reverse the weighting function in time, the resulting function represents the effect of the current point on the output of the filter. This reversed weighting function is equivalent to the waveform that the filter would produce in response to a momentary voltage spike or *impulse*; it is therefore known as the *impulse response function* of the filter.

Figure 5.5 illustrates the relationship between the weighting function and the impulse response function. Panel A shows a weighting function, and panel B shows the corresponding impulse response function. In addition, panel C shows the output of the filter if the input is a momentary impulse. Note that the output of the filter becomes non-zero before the impulse, which is possible only with an off-line, digital filter. Analog filters that operate in real time have impulse response functions that are zero before the time of the impulse.

Although it may seem more natural to think of filtering in terms of the original weighting function, most mathematical descriptions of filtering rely instead on the impulse response function. One reason for this is that it is possible to measure the impulse response function of an analog filter empirically by providing an impulse for an input and simply measuring the output of the filter. Of course, the impulse response function is identical to the weighting function if the function is symmetrical about time zero, which is usually the case with digital filters, making the two conceptualizations of filtering identical.

To use the impulse response function instead of the original weighting function for computing the filtered value at each time point in the waveform, it is necessary to make a small change to the formula presented in equation 5.2:

$$fERP_i = \sum_{j=-n}^{n} IRF_j ERP_{i-j} \tag{5.3}$$

where IRF_j is the value of the impulse response function at time j, which is the same as the original weighting function at time $-j$. This equation essentially reverses the coefficients of the impulse response function—thus creating our original weighting function—and then performs the same filtering operation described in equation 5.2.

When expressed in this manner, the combination of the impulse response function and the ERP waveform is termed a *convolution*, which is typically symbolized in mathematical equations by the $*$ operator. We can therefore write equation 5.3 as:

$$fERP = IRF * ERP \tag{5.4}$$

This equation states that the filtered ERP waveform is equal to the convolution of the impulse response function and the unfiltered ERP waveform (note that the $*$ symbol is often used to denote multiplication, especially in computer languages; we will use $\times$ to denote multiplication and $*$ to denote convolution).

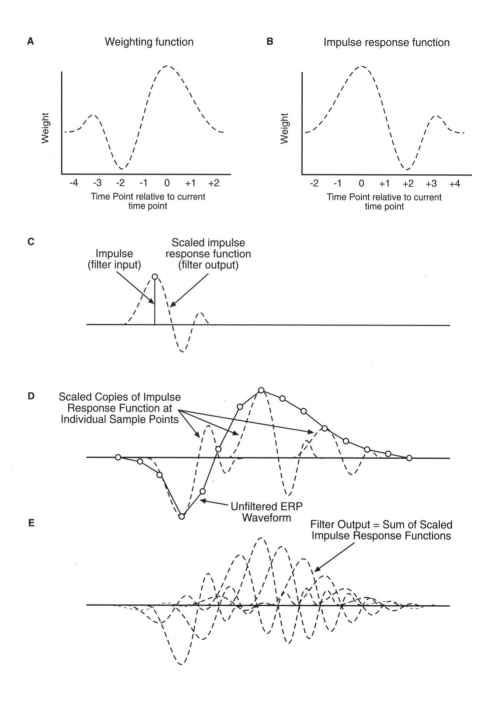

The filtering equations listed above are written in a manner that makes it easy to see how the filtered value at a given time point is computed from the unfiltered values at surrounding time points. It is also possible to take a complementary approach and compute the contribution of a given unfiltered time point to the filtered waveform. Although less useful computationally, this approach more readily allows one to visualize the relationship between a filter's impulse response function and the filtered ERP waveform. In this approach, each point in the unfiltered waveform is replaced by a scaled copy of the impulse response function (scaled by the amplitude of the unfiltered waveform at the current time point). These scaled copies of the impulse response function are then simply added together to compute the filtered ERP waveform.

Figure 5.5 illustrates this approach to filtering. This figure shows an arbitrary weighting function (panel A), which was designed solely for the purposes of illustration. The impulse response function of the filter (panel B) is equal to the weighting function reflected around time zero. If the input to the filter is a brief impulse, as shown in panel C, then the output of the filter is simply a copy of the impulse response function scaled by the size of the impulse (by definition). Panel D shows an unfiltered ERP waveform that was sampled at each of the points indicated by open circles (the lines connecting the circles are interpolations). This

◀ **Figure 5.5** Filtering by convolving an ERP waveform with the filter's impulse response function. The weighting function of the filter (A) is reversed in time to produce the filter's impulse response function (B). As shown in (C), the impulse response function is the output of the filter in response to a brief impulse. ERP waveforms are analogously filtered by treating each sample point as an impulse and replacing each sample point with a scaled copy of the impulse response function, as shown in (D) and (E). In (D), the unfiltered ERP waveform is represented by the solid waveform, and each open circle represents a sample point; to illustrate the scaling process, this panel also shows the scaled impulse response functions corresponding to three of the sample points. Note that the function is inverted when the ERP voltage is negative. (E) Shows the result of replacing every point in the ERP waveform with a scaled copy of the impulse response function; the filtered waveform is equal to the sum of these scaled impulse response functions.

waveform is filtered by replacing each data point with a scaled copy of the impulse response function (for the sake of simplicity, panel D shows this for only three data points). Panel E shows all of the scaled copies of the impulse response function, which are added together to compute the filtered ERP waveform (not shown). By conceiving of filtering in this manner, it is possible to visualize how a filter's output is related to its impulse response function, as discussed further below.

It may seem strange that filtering can be achieved by simply replacing each data point with the impulse response function and summing. In fact, this is true only for a subset of filters, called *finite impulse response* filters. For these filters, the filter's output does not feed back into its input, which leads to this simple pattern of behavior. It is possible to design more sophisticated filters that do not obey this rule (called *infinite impulse response* or *recursive* filters), but these filters are unnecessarily complex for the needs of most ERP experiments.

Relationship Between the Time and Frequency Domains

We have now seen how filtering can be accomplished in the frequency domain and in the time domain. These may seem like completely different procedures, but there is a fundamental relationship between the time-domain and frequency-domain approaches to filtering that allows a straightforward conversion between a filter's impulse response function and its transfer function. This relationship is based on an important mathematical principle: *Multiplication in the frequency domain is equivalent to convolution in the time domain.* As a result of this principle, convolving an ERP waveform with an impulse response function in the time domain is equivalent to multiplying the frequency-domain representation of the ERP waveform with the frequency-domain representation of the impulse response function.

This implies an important fact about filters, namely that the Fourier transform of a filter's impulse response function is equal to the filter's transfer function (remember that the transfer function is the combination of a frequency response function and a phase response function). It is therefore possible to determine a filter's transfer function by simply transforming its impulse response function into the frequency domain by means of the Fourier transform. Conversely, the inverse Fourier transform of a transfer function is equal to the impulse response function of the filter. As a result, one can compute the impulse response function that will yield a desired transfer function by simply transforming the transfer function into the time domain. Figure 5.6 illustrates this.

It is usually substantially faster to filter via convolutions than via Fourier transforms, and because the two methods yield identical results, most digital filters are implemented by means of

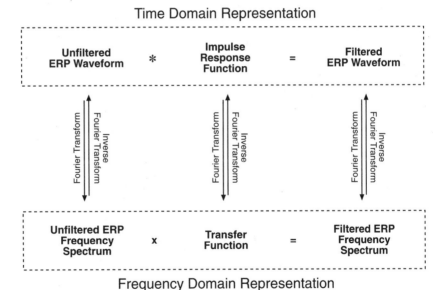

Figure 5.6   Relationship between filtering in the time and frequency domains, showing: (1) that each term in the time domain can be converted to a corresponding term in the frequency domain by means of the Fourier transform; and (2) that convolution (represented by the ∗ operator) in the time domain is equivalent to multiplication (represented by the × operator) in the frequency domain.

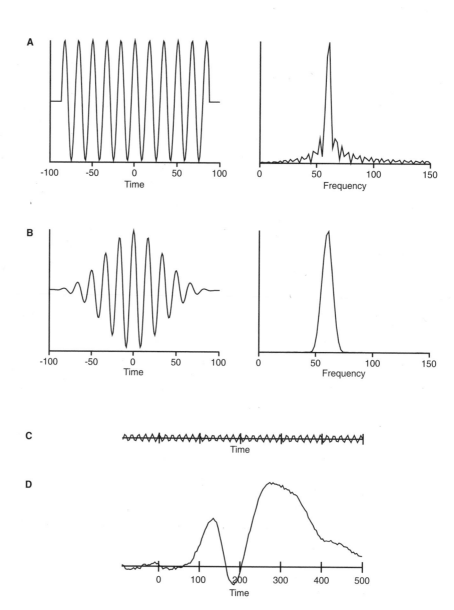

Figure 5.7   Relationship between the time and frequency domains for 60-Hz filtering. (A) Time- and frequency-domain representations of a finite-duration 60-Hz sine wave. (B) Time- and frequency-domain representations of a windowed sine wave (tapered with a Blackman window). (C) Result of convolving the 60-Hz sine wave in (A) with the ERP waveform shown in figure 5.1A, which extracts the 60-Hz component of

convolutions. The method usually recommended for constructing filters is therefore to create the desired transfer function for the filter and then transform this function into the time domain to create an impulse response function; this impulse response function can then be convolved with the ERP waveform to create the filtered ERP waveform.[4] However, although this approach yields very nice results in the frequency domain, it may lead to significant distortions in the time domain, as I will describe below.

As an example of the relationship between the impulse response function and the corresponding frequency response function, consider how a filter would be constructed to pass the 60-Hz frequency band in an ERP waveform and eliminate all other frequencies. The frequency response function of such a filter would have a magnitude of 1.0 at 60 Hz and a magnitude of 0.0 at all other frequencies. Transferring this function into the time domain to derive the impulse response function of the filter would simply yield a sine wave at 60 Hz, and convolving a 60-Hz sine wave with an ERP waveform would therefore extract the 60-Hz component from the waveform.

In practice, filtering everything except the 60-Hz activity would be complicated by the fact that the impulse response function must be finite in duration, as figure 5.7 illustrates. Panel A of the figure shows how a 60-Hz sinusoidal impulse response function of finite length yields a frequency response function that contains a peak at 60 Hz but also contains power at a broad set of higher and lower frequencies. The power at these other frequencies is due to the sudden transitions at the beginning and end of the impulse response function, which is unavoidable with a sine wave of finite duration. As panel B shows, you can mitigate this problem by multiplying the sinusoidal impulse response function by a windowing

◀ **Figure 5.7**   (continued)
the waveform. (D) Effect of removing the 60-Hz component shown in (C) from the unfiltered ERP waveform, which effectively eliminates the 60-Hz artifact from the waveform without attenuating the higher or lower frequencies.

function to taper the ends of the function. This yields a frequency response function that, although somewhat broader around 60 Hz, is smoother and falls to zero sooner than the frequency response function of the pure sinusoid. Panel C shows the results of convolving this windowed sinusoid with the ERP waveform shown in figure 5.3A, which effectively extracts the 60-Hz component from the ERP waveform. Panel D of figure 5.7 shows how subtracting this 60-Hz component from the original waveform provides a means of eliminating the 60-Hz noise from the waveform. Unlike the filter used in figure 5.3E to attenuate all high frequencies, this filter has removed only frequencies around 60 Hz, allowing information at both lower and higher frequencies to remain in the filtered waveform.

Time-Domain Distortions Produced by Filters

In the above examples, we have seen how noise can be attenuated from ERP waveforms by means of filtering. However, transient ERP waveforms necessarily contain a broad range of frequencies, and filtering a restricted range of frequencies from a waveform consisting of both noise and an ERP response will attenuate those frequencies from both the noise and the ERP signal. This will almost always lead to some distortion of the time-domain ERP waveform, and the distortion may range from mild to severe depending on the nature of the impulse response function and the ERP waveform.

Distortions Produced by Notch Filters

To illustrate the distortion produced by filtering, figure 5.8 shows the effects of applying several types of notch filters to an artificial ERP waveform consisting of one cycle of a 10-Hz sine wave. Panel A of this figure shows the artificial ERP waveform and its Fourier transform. Note that although the ERP waveform consists of a portion of a 10-Hz sine wave, it contains a broad range of frequencies because it is restricted to a narrow time interval (remember, a narrow distribution over time corresponds to a broad distribution over frequencies, and vice versa). If we apply a filter to remove the 10-Hz component, therefore, we will not eliminate the entire ERP waveform, but only the 10-Hz portion of it. This is illustrated in figure 5.8B, which shows the impulse response function for a 10-Hz notch filter, the filter's frequency response function, and the results of applying this filter to the waveform shown in figure 5.8A.

The impulse response function was created by windowing a 10-Hz sine wave, as in figure 5.7B, and then converting it into a filter that removes rather than passes power at 10 Hz (see below). The resulting impulse response function has a positive peak at time zero, surrounded on both sides by oscillations at ~10 Hz that are 180 degrees out of phase with the artificial ERP waveform; these opposite-phase oscillations cause power at ~10 Hz to be subtracted away from the original ERP waveform. When applied to the artificial ERP waveform, the peak amplitude of the waveform is therefore reduced, but the oscillation in the impulse response function is carried over into the filtered waveform, where it can be seen both in the prestimulus interval and in the time period following the end of the original response. The frequency spectrum of the filtered ERP waveform has a sharp drop to zero power at 10 Hz, but the nearby frequencies still have significant power, and these nearby frequencies are the source of the oscillation that can be observed in the filtered waveform. Thus, a filter designed to eliminate the 10-Hz component from a waveform can actually produce an output containing artificial oscillations near 10 Hz that weren't present in the input. Although these oscillations are more extreme than those that a more typical combinations of filters and ERP waveforms would produce, this example demonstrates that impulse response functions that contain oscillations can induce artificial oscillations in the filtered ERP waveform. This important fact is often overlooked when filtering is considered as a frequency-domain operation rather than a time-domain operation.

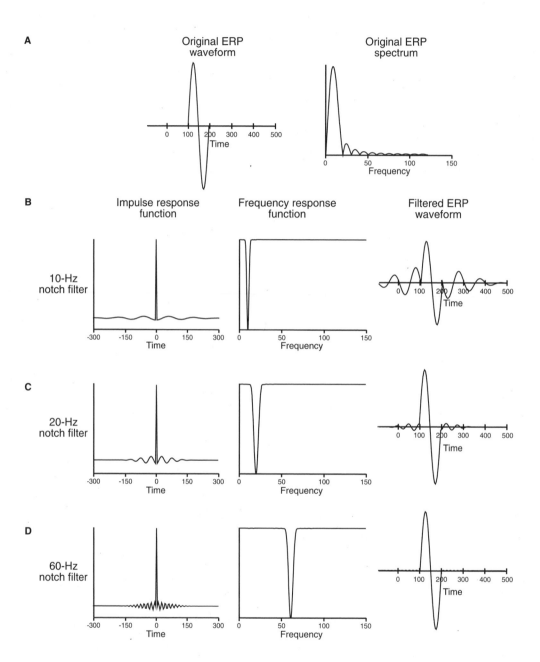

A

Original ERP
waveform

Original ERP
spectrum

B

Impulse response
function

Frequency response
function

Filtered ERP
waveform

10-Hz
notch filter

C

20-Hz
notch filter

D

60-Hz
notch filter

The presence of oscillations in the impulse response function is not alone sufficient to produce large oscillations in the output of the filter. As figures 5.8C and 5.8D show, notch filters at 20 Hz and 60 Hz produce much smaller oscillations than the 10-Hz notch filter in the context of this particular artificial ERP waveform. This is due to the fact that there is much less power at these frequencies in the spectrum of the unfiltered ERP waveform (see figure 5.8A). Thus, you must consider both the impulse response function and the nature of the unfiltered ERP waveform to determine the distortion that a filter will produce.

From these filter-induced distortions, it should be clear that you must know the shape of the impulse response function to assess the distortions that a filter might produce. For example, a filter that is simply labeled "60-Hz notch filter" on an EEG amplifier will have a very different impulse function from the filter shown in figure 5.8D, and may lead to much greater distortion of the ERP waveform than the minimal distortion present in figure 5.8D. Thus, the common practice of specifying only the half-amplitude or half-power cutoff of a filter is clearly insufficient; descriptions of filtering should specify the impulse response function in addition to the cutoff of the filter. For example, when I describe a filter in a journal article, I write something like this: "The ERP waveforms were low-pass filtered offline by convolving them with a Gaussian impulse response function with a standard deviation of 6 ms and a half-amplitude cutoff at ~30 Hz."

Distortions Produced by Low-Pass Filters

In many cases, it is useful to attenuate all frequencies above some specified point (low-pass filtering). Such filtering is necessary

◄ **Figure 5.8**   Examples of temporal distortions produced by notch filters. (A) Shows the original artificial ERP waveform (one cycle of a 10-Hz sine wave) and its Fourier transform. (B), (C), and (D) show the impulse response and frequency response functions for 10-, 20-, and 60-Hz notch filters and also display the result of applying these filters to the waveform shown in (A).

during digitization of the raw EEG data, because you must sample at a rate that is at least twice as high as the highest frequency in the incoming data. Low-pass filtering is also useful for attenuating the relatively high-frequency noise caused by muscle activity or by external electrical devices (e.g., line-frequency noise at 50 or 60 Hz). The cutoff frequency used for a given experiment will depend on the frequency content of the ERPs being recorded and the frequency content of the noise to be filtered. For example, brainstem evoked responses contain substantial power at very high frequencies, making a 5-KHz cutoff appropriate, but this makes it difficult to filter out muscle noise, which falls into the same frequency range. In contrast, the long-latency ERP components consist primarily of power under about 30 Hz, making a 30–100 Hz cutoff appropriate and allowing high-frequency muscle activity to be filtered without much distortion of the underlying ERP waveform. However, there is almost always some overlap between the frequencies in the ERP waveform and in the noise, and some filter-induced distortion is therefore inevitable.

This leads us once again to Hansen's axiom: *There is no substitute for good data.* Filters necessarily cause distortion, and it is always better to reduce noise at the source (e.g., through shielding, low electrode impedance, high common-mode rejection, careful experimental design, etc.) than to attenuate it with filters. When you cannot avoid filtering, it is important to choose a filter with an optimal balance between suppression of noise and distortion of the ERP waveform. Depending on the nature of the experimental question being addressed, some types of distortion may be more problematic than others, and there are therefore no simple rules for designing an optimal filter.

Most discussions of filter design emphasize the frequency domain and therefore lead to filters such as the *windowed ideal* filter shown in figure 5.9A. This particular windowed ideal filter has a half-amplitude cutoff at 12.5 Hz and is an "ideal" filter because it perfectly passes all frequencies below approximately 12 Hz and completely suppresses all frequencies above approximately 13 Hz, with only a narrow "transition band" in which the attenuation is incomplete. This filter also produces no phase distortion because its impulse response function is symmetric around time zero. Despite the usefulness of these attributes, the sharp transitions in the frequency response function of this filter require a broad, oscillating impulse response function, which leads to substantial distortion in the time domain. This can be seen when this filter is applied to the artificial ERP waveform from figure 5.8A, yielding an output containing damped oscillations both before and after the time range of the original ERP waveform[5] (see figure 5.9A, right column). Importantly, the filtered waveform contains peaks at around 70 ms and 230 ms that are completely artifactual and are a consequence of the oscillations in the filter's impulse response function. Thus, filters with sharp cutoffs in their frequency response functions may lead to large distortions in the apparent onset and offset of an ERP waveform and to artifactual peaks that are not present in the original ERP waveform. In other words, a filter that is ideal in terms of its frequency-domain properties may be far from ideal in terms of its time-domain properties.

Let us consider briefly how this type of distortion might influence the interpretation of an ERP experiment. As an example, consider the ERP waveform elicited by a visual stimulus over frontal cortex, which typically contains an N1 component at about 130 ms but has no prior peaks (the earlier P1 component is typically absent at frontal sites). If this waveform were filtered with the low-pass filter shown in figure 5.9A, the filtered waveform would contain an artifactual positive peak preceding the N1 component and peaking at about 70 ms post-stimulus, and this artifactual peak might be mistaken for the P1 component. Moreover, if two conditions were compared, one of which produced a larger N1 component than the other, these amplitude differences would also be observed in the artifactual pseudo-P1 component. On the basis of the filtered response, one might conclude that the experimental manipulation caused an increase in P1 amplitude at 70 ms even though there was no real P1 component and the experimental

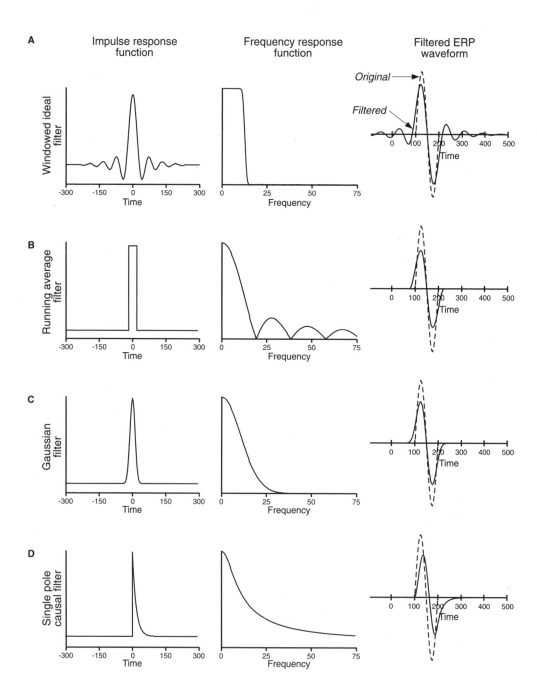

A   Impulse response
function

Frequency response
function

Filtered ERP
waveform

Windowed ideal
filter

Original

Filtered

-300   -150   0   150   300
Time

0   25   50   75
Frequency

0   100   200   300   400   500
Time

B   Running average
filter

-300   -150   0   150   300
Time

0   25   50   75
Frequency

0   100   200   300   400   500
Time

C   Gaussian
filter

-300   -150   0   150   300
Time

0   25   50   75
Frequency

0   100   200   300   400   500
Time

D   Single pole
causal filter

-300   -150   0   150   300
Time

0   25   50   75
Frequency

0   100   200   300   400   500
Time

effect did not actually begin until 100 ms. This example underscores the necessity of knowing the impulse response function of a filter to avoid making incorrect conclusions about the time course of ERP activity.

The bottom three panels of figure 5.9 show several alternative filters that have approximately the same 12.5-Hz half-amplitude cutoff frequency as the windowed ideal filter shown in the top panel, but produce substantially less waveform distortion. The first of these is a simple running average filter that is equivalent to averaging successive points together (as described in the initial discussion of low-pass filtering). The impulse response function of this filter extends over a very narrow time range compared to the windowed ideal filter, and therefore causes much less temporal spread in the filtered ERP waveform. As the right column of figure 5.9B shows, the output of the running average filter begins slightly before and ends slightly after the original waveform, but the filter causes relatively little change in the overall shape of the waveform. This filter has two shortcomings, however. First, there is substantial attenuation in the 10-Hz frequency band where most of the power of the original ERP waveform is located, so the filtered waveform is somewhat smaller than the original. Second, the frequency response function of the filter does not fall monotonically to zero. Instead, there are *side lobes* that allow substantial noise from some high frequencies to remain in the filtered waveform (this cannot be seen in the examples shown in figure 5.9, which contain no high-frequency noise). These side lobes can be predicted from the square shape of the impulse response function: because the transfer function is the Fourier transform of the impulse response function, the high frequencies required for the

◀ **Figure 5.9**   Examples of temporal distortions produced by several types of low-pass filters, each with a half-amplitude cutoff of approximately 12.5 Hz. The left and middle columns show the impulse and frequency response functions of the filters, respectively. The right column shows the result of applying each filter to the artificial ERP waveform shown in figure 5.8A.

sudden onset and offset of the impulse response function lead to the presence of substantial high frequency power in the frequency response function.

Figure 5.9C shows a filter with a gaussian impulse response function that has approximately the same temporal spread as the running average impulse response function. Like the running average filter, the gaussian filter produces some smearing in the onset and offset of the filtered ERP waveform and some attenuation of the overall waveform. However, the frequency response function of the gaussian filter falls monotonically to zero, leading to virtually complete attenuation of frequencies greater than 30 Hz. In most cases, a gaussian filter provides the best compromise between the time and frequency domains, with a monotonic and fairly rapid fall-off in the frequency response function combined with minimal temporal distortion of the ERP waveform. Typically, a gaussian filter with a half-amplitude cutoff of 30 Hz will eliminate most line-frequency and muscle noise while producing very little attenuation of the long-latency ERP components and relatively little latency smearing. In the vast majority of cases, I would recommend using gaussian impulse response functions for low-pass filtering.

The frequency-domain properties of filters are often described by a single number, the half-amplitude cutoff frequency. The time-domain properties of a gaussian filter can also be described with a single number that reflects the width of the impulse response function. There are two common ways to do this. First, you can indicate the standard deviation of the gaussian. Second, you can indicate how many milliseconds wide the gaussian function is at half of its maximum value, which is called the *full width at half maximum* (FWHM). fMRI experiments commonly use spatial gaussian filters, and FWHM is the usual way of describing the impulse response function in that context. These values can be interconverted with the filter's half-amplitude cutoff. The half-amplitude cutoff in Hz of a gaussian filter with a standard deviation of σ milliseconds is simply $185.5/\sigma$ (e.g., a gaussian filter with a standard deviation of 4 ms would have a half amplitude cutoff of 185.5/4

Hz). The FWHM in milliseconds is equal to 2.355 times the standard deviation in milliseconds, and the half-amplitude cutoff in Hz can be computed as 79.62/FWHM.

The windowed ideal, running average, and gaussian filters described so far are among the most commonly used low-pass digital filters for ERP research, and each has certain advantages and disadvantages. Because of its sharp cutoff, the windowed ideal function may be appropriate when the ERP waveform and the to-be-filtered noise contain substantial power in nearby frequency ranges. However, this filter type leads to substantial distortion in the time domain, which may lead to incorrect conclusions about the timing of an ERP component and may even lead to spurious peaks before the onset or after the offset of the real ERP waveform. In recordings of long-latency ERP components, the noise is usually concentrated at substantially higher frequencies than the majority of the ERP power, making the sharp cutoff of the windowed ideal filter unnecessary, and a gaussian or running average filter is therefore more appropriate. One exception to this occurs when an ERP waveform is contaminated by alpha-frequency ($\sim$ 10 Hz) EEG noise that interferes with late components such as the P300, which have their power concentrated at slightly lower frequencies. The best way to eliminate alpha noise is usually to maintain subject alertness and average together a large number of trials (consistent with Hansen's axiom), but this is not always feasible. When necessary, it is possible to use a windowed ideal filter with a half-amplitude cutoff frequency of around 8 Hz to attenuate alpha activity, but the resulting data must be interpreted cautiously because of the time-domain distortions that such a filter will inevitably produce.

The running average and gaussian filters produce similar patterns of temporal distortion, characterized primarily by the smearing of onset and offset times, but the gaussian filter is usually preferable because its frequency response function falls to zero at high frequencies. The running average filter is somewhat easier to implement, however, which is important in some applications. In addition, it is sometimes possible to take advantage of the multiple zero points in the frequency response function of the running average filter. If the primary goal of filtering is to attenuate line-frequency noise, it may be possible to employ a running average filter where the zero points fall exactly at the line frequency and its harmonics, producing excellent attenuation of line frequency noise.

The last class of low-pass filters I will discuss here are called *causal* filters and are used in analog filtering devices such as EEG amplifiers (but can also be implemented digitally). These filters are labeled *causal* because they reflect the normal pattern of causation in which an event at a particular time can influence subsequent events but not previous events. More precisely, the impulse response functions of causal filters have values of zero for times preceding time zero. Viewed from the opposite temporal perspective, the output of the filter at a given time point reflects only the input values at previous time points and not the input values at subsequent time points. Figure 5.9D shows an example of such a filter, displaying the impulse response function of a very simple analog filter. Filters that do not obey the normal pattern of causation (i.e., because they nonzero values before time zero in their impulse response functions) are called *noncausal* filters.

Because their impulse response functions extend only one direction in time, causal filters produce shifts in peak latency and in offset time, which are typically considered negative traits. However, causal filters may produce relatively little distortion of onset latency, which may provide an advantage over noncausal filters when onset latency is an important variable (see Woldorff, 1993). This is not always true of causal filters, however; it depends on a relatively rapid, monotonically decreasing fall-off from time zero in the impulse response function, as in the filter shown in figure 5.9D.

The frequency response function of the causal filter shown in figure 5.9D is suboptimal for most purposes because it falls relatively slowly and never reaches zero. This is not an inherent property of causal filters, however. For example, if the gaussian impulse response function shown in figure 5.9C were simply shifted to the right so that all values were zero prior to time zero, the resulting causal filter would have the same frequency response function as the noncausal gaussian filter, but the output waveform would be shifted in time by the same amount as the shift in the impulse response function. Bessel filters, which can be implemented in analog circuitry, have an impulse response function that approximates a shifted gaussian and therefore make an excellent choice for on-line filtering (e.g., during EEG digitization). Bessel filters also have a linear phase response, which means that all frequencies in the output waveform are shifted by the same amount of time. This time shift can be computed and the filtered waveform can be shifted backwards by this amount off-line, thus eliminating the latency shifts inherent in analog filters. Unfortunately, Bessel filters are expensive to implement in analog circuitry, so they are relatively uncommon.

Some Important Properties of Convolution

Before going further, it is useful to discuss some important mathematical properties of the convolution operation. In particular, convolution has the same commutative, associative, and distributive properties as multiplication. The commutative property states that it doesn't matter which of two functions comes first in a convolution formula. More precisely:

$$A * B = B * A$$

The associative property states that multiple consecutive convolutions can be performed in any order. More precisely:

$$A * (B * C) = (A * B) * C$$

The distributive property states that the convolution of some function A with the sum of two other functions B and C is equal to A convolved with B plus A convolved with C. More precisely:

$$A * (B + C) = (A * B) + (A * C)$$

These mathematical properties of convolution lead to several important properties of filters. For example, one common question about filtering is: what happens when you filter a filtered waveform a second time? If we have two filters with impulse response functions denoted by IRF_1 and IRF_2 and an ERP waveform denoted by ERP, we can write the process of filtering twice as:

$$(ERP * IRF_1) * IRF_2$$

Because of the associative property, this is equal to convolving the two impulse response functions first and then convolving the result with the ERP waveform:

$$ERP * (IRF_1 * IRF_2)$$

To take a concrete example, the convolution of two gaussian functions yields a somewhat wider gaussian, and filtering an ERP twice with a gaussian filter is therefore equivalent to filtering once with a wider gaussian.

In addition, because convolution in the time domain is equivalent to multiplication in the frequency domain, the frequency response function of the double filtering is equal to the product of the frequency response functions of the two individual filters. In the case of two gaussian filters, this would lead to greater attenuation of high frequencies than either filter would produce alone. In the more common case of the application of both a low-pass filter and a high-pass filter, the result is a band-pass filter that simply cuts both the high and low frequencies. However, if both filters have relatively gradual frequency response functions, the multiplication of these functions may also lead to fairly substantial attenuation of the intermediate frequencies.

Time-Domain Implementation of High-Pass Filters

In the initial discussion of high-pass filters near the beginning of the chapter, I mentioned that one can accomplish high-pass filtering by creating a low-pass-filtered waveform and subtracting this from the unfiltered waveform, thus yielding the high frequencies that are present in the unfiltered waveform and absent in the low-pass-filtered waveform. This process can be expressed as:

$$ERP_H = ERP - ERP_L = ERP - (IRF_L * ERP)$$

where ERP_H and ERP_L are the high- and low-pass-filtered ERP waveforms and IRF_L is the impulse response function of the low-pass filter. This convolution-and-subtraction sequence can be replaced by a single convolution with a high-pass impulse response function that can be computed by some simple algebraic manipulations. First, it is necessary to define a "unity" impulse response function IRF_U that is 1.0 at time zero and 0.0 at all other points; the convolution of this function with an ERP waveform would be equal to the ERP waveform (analogous to the multiplication of a number by 1.0). By using this unity function, we can use the distributive property to create an impulse response function IRFH that will accomplish high-pass filtering in a single step:

$$ERP_H = ERP - (IRF_L * ERP)$$

$$= (IRF_U * ERP) - (IRF_L * ERP) \quad \text{[because ERP = IRF}_U * \text{ERP]}$$

$$= (IRF_U - IRF_L) * ERP \quad \text{[because of the distributive property]}$$

$$= IRF_H * ERP, \quad \text{where } IRF_H = IRF_U - IRF_L$$

Thus, we can create a high-pass impulse response function by subtracting a low-pass impulse response function from the unity impulse response function ($IRF_U - IRF_L$). The frequency response function of the resulting high-pass filter is simply the complement of the frequency response function of the low-pass filter (1.0 minus the response at each individual frequency point in the low-pass frequency response function).

Figure 5.10A diagrams the creation of a high-pass impulse response function from a unity impulse response function and a gaussian low-pass impulse response function with a half-amplitude cutoff of 2.5 Hz. The unity impulse response function (figure 5.10A, left) consists of a single spike of magnitude 1.0 at

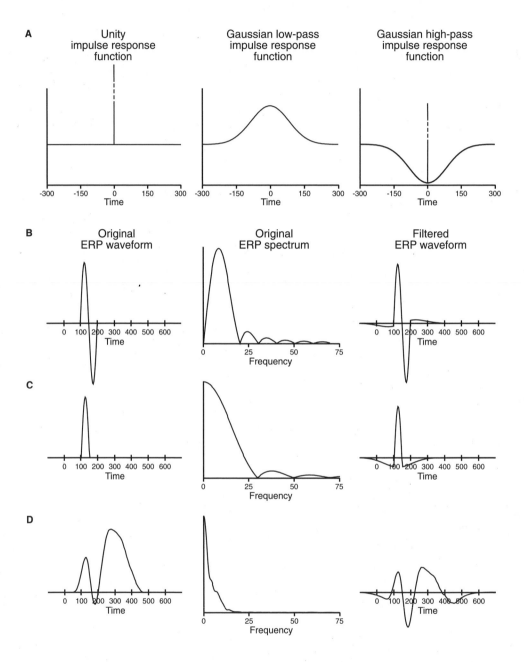

time zero, which is very large compared to the values at individual time points in the gaussian low-pass impulse response function (figure 5.10A, middle). The high-pass impulse response function is created by simply subtracting the low-pass impulse response function from the unity function, as shown at the right of figure 5.10A.

A related attribute of the class of filters discussed here is that they are linear, so you can combine filtering with other linear operations in any order. For example, signal averaging is also a linear process, and this means that filtering an averaged ERP waveform yields the same result as filtering the EEG data before averaging.[6] Averaged ERP data sets are typically considerably smaller than the EEG data from which they are derived, and filtering after averaging is therefore more efficient than filtering the raw EEG and then averaging.

Distortions Produced by High-Pass Filters

Figure 5.10 shows the application of a gaussian high-pass filter to two artificial ERP waveforms and one realistic ERP waveform. The inverted gaussian in the impulse response function is visible in the filtered ERP waveforms, where it produces overshoots that one can observe at the beginning and end of the waveforms. These distortions are particularly evident for ERP waveforms that consist primarily of one polarity, such as those in figures 5.10C and 5.10D. This can be understood by viewing filtering as replacing each sample in the ERP waveform with a scaled copy of the impulse response function: when the unfiltered ERP waveform consists of roughly equivalent positive and negative subcomponents, as in figure 5.10B, these subcomponents will lead to opposite-polarity copies of the impulse response function, which will cancel each

◀ **Figure 5.10** Construction and application of a gaussian high-pass filter. (A) A unity impulse response function multiplied by a gaussian low-pass impulse response function yields a gaussian high-pass impulse response function. (B), (C), and (D) show the application of this filter to three different ERP waveforms.

other out to some extent. The amount of cancellation will depend on the exact shapes of the waveform and the impulse response function, but in most cases the overshoot will be greater when one polarity dominates the ERP waveform.

The distortions in figure 5.10 are in some sense similar to the distortions the gaussian low-pass filter shown in figure 5.9C produced, but reversed in polarity because the gaussian in the high-pass impulse response function is inverted (because of the lower cutoff frequency of the high-pass filter, the distortion in figure 5.10 is also somewhat broader than the distortion in figure 5.9C). Thus, when an ERP waveform is low-pass filtered, the gaussian impulse response function produces a spreading of the peaks that is the same polarity as the peaks, whereas using a high-pass filter produces opposite-polarity spreading. Although these distortions are similar in many ways, the distortions the high-pass filter produces may be more problematic because they may lead to the appearance of artifactual peaks, such as the peaks at approximately 75 ms and 450 ms in the filtered waveform shown in figure 5.10D.

Figure 5.11 shows the effects of several different types of high-pass filters (all with half-amplitude cutoffs at approximately 2.5 Hz) on a realistic ERP waveform. Although the impulse response and frequency response functions of the windowed ideal and gaussian filters (panels A and B) look very similar, the windowed ideal impulse response function contains damped oscillations just like those in the windowed ideal low-pass filter in figure 5.9A. The half-amplitude cutoff frequency of the high-pass filter shown here is so low, however, that these oscillations fall outside of the plotted time window; with a higher cutoff frequency, these oscillations would be more apparent. Thus, windowed ideal high-pass filters may induce artificial oscillations in the filtered ERP waveform, whereas gaussian high-pass filters do not. As discussed above and shown in figure 5.11B, however, gaussian high-pass filters can produce individual artifactual peaks at the beginning and end of the filtered waveform, unlike gaussian low-pass filters. For this reason, you must take extreme caution in interpreting high-pass-

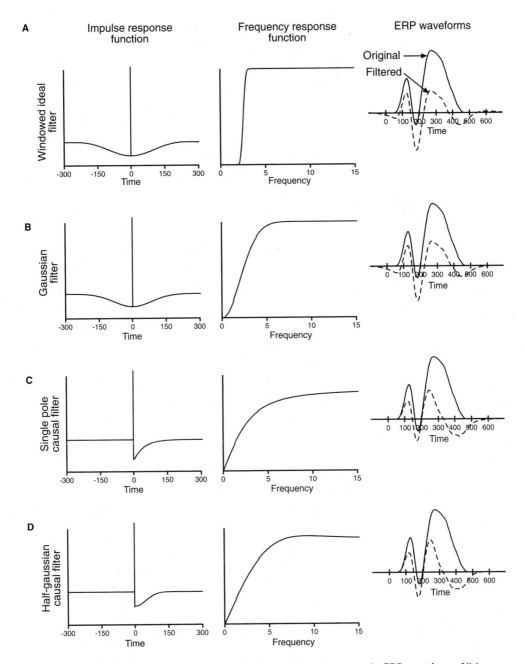

Figure 5.11 Application of four different high-pass filters to a sample ERP waveform. All have a 50 percent amplitude cutoff near 2.5 Hz.

filtered waveforms, even when using a gaussian impulse response function.

Figure 5.11C shows the impulse response and frequency response functions of the type of causal filter found in older EEG amplifiers. Because its impulse response function is zero for all points before time zero, this filter cannot produce any distortion before the onset of the unfiltered ERP waveform and thus does not produce an artifactual peak at the beginning of the filtered ERP waveform. However, this same attribute of causal filters leads to distorted peak latencies in the filtered waveform, whereas this form of distortion is absent for most noncausal filters. These different forms of distortion underscore the important principle that all filters produce distortions, and the choice of which filter to use will depend upon the goals of the experiment and the types of distortion that are acceptable. The filter shown in figure 5.11C also has a relatively gradual roll-off; this can be improved by using the filter in figure 5.11D, which uses the right half of a gaussian in its impulse response function. Like the filter in figure 5.11C, this "half-gaussian" filter produces minimal distortion at the beginning of the filter ERP waveform, but has a somewhat sharper roll-off and is therefore a better choice in many cases.

In most cases, noncausal high-pass filters tend to take low-frequency information away from the time zone of the unfiltered ERP waveform and "push" it forward and backward in time in the filtered ERP waveform. In contrast, causal high-pass filters of the type shown in figure 5.11 push the information exclusively to later time points. Figure 5.12 illustrates this. ERP waveforms typically consist of small, relatively high-frequency early components followed by larger, relatively low-frequency late components, and this asymmetrical sequence makes the bidirectional distortions produced by noncausal high-pass filters particularly harmful. In particular, low frequency information from the large, low-frequency later components is pushed into the latency range of the smaller early components, causing substantial distortion of the early components even though they contain relatively little low-

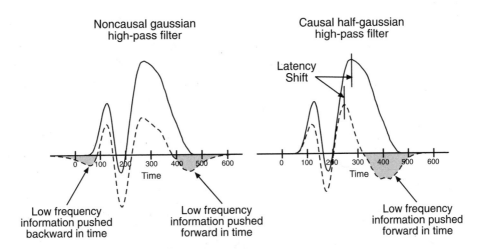

Figure 5.12   Comparison of a noncausal gaussian high-pass filter and a causal half-gaussian high-pass filter. Note that the noncausal filter produces obvious distortions at the beginning and end of the waveform, whereas the causal filter produces no distortion at the beginning of the waveform. Note also that the causal filter produces a large latency shift, but the noncausal filter does not.

frequency information. Using a causal filter such as those shown in panels C and D of figure 5.11, however, will minimize distortion of the early components because low-frequency information from the late components will not be pushed backward into the time range of the early components. Note, however, that this is true only for filters that have monotonically increasing impulse response functions after time zero; filters that do not possess this property, such as Bessel filters, may cause distortions similar to those produced by noncausal filters. In addition, causal filters may produce substantial latency shifts that may be problematic in some cases.

Recommendations Revisited

I discussed some recommendations about filtering early in this chapter. Now that you have a deeper understanding of how filters actually work, you might want to read through those recommendations again.

My two most important recommendations are as follows. First, filter the data as little as possible. As you've seen, filters always distort the time-domain properties of ERP waveforms, and you don't want to distort your waveforms unnecessarily. My second main recommendation is that before you use a filter, think about exactly how it might distort your data. Often, the best way to see how a filter will distort your data is to create a simulated, noise-free waveform and apply the filter to it. This will allow you to see how the filter distorts component onsets and offsets and adds spurious components or oscillations to your data. If the effects of the filter on the simulated data aren't too bad, then you can use the filter on your real data with confidence that the filter will help you rather than hurt you.

The previous chapters have discussed designing experiments, collecting the data, and applying signal processing procedures; this chapter discusses the final steps in an ERP experiment: plotting the data, measuring the components, and subjecting these measurements to statistical analyses.

Plotting ERP Data

Although it might seem trivial to plot ERP waveforms, I regularly read ERP papers in which the waveforms are plotted in a way that makes it difficult to perceive the key aspects of the waveforms. Thus, I will begin this chapter with a few recommendations about plotting ERP data.

The most important thing to keep in mind is that virtually all ERP papers should include plots of the key ERP waveforms. Given all of the difficulties involved in isolating specific ERP components, it is absolutely necessary to see the waveforms before accepting the validity of a paper's conclusions. Remember that Kramer (1985) showed that ERP experts do a good job of determining the underlying latent components when they see the observed ERP waveforms, and presenting the waveforms is essential for this. In fact, the official publication guidelines of the Society for Psychophysiological Research (SPR) state that "the presentation of averaged ERP waveforms that illustrate the principal phenomena being reported is mandatory" (Picton et al., 2000, p. 139).

The SPR guidelines describe several additional elements of ERP figures that you should keep in mind. First, it is almost always a good idea to show the data from multiple electrode sites, spanning the region of the scalp where the effects of interest were present. This information plays a key role in allowing experts to determine the underlying component structure of the waveform. However, including multiple electrode sites is not as important when you can isolate components by other means (e.g., by performing the subtraction procedure that isolates the lateralized readiness potential).

Second, plots of ERP waveforms should indicate the voltage scale and the time scale in a way that makes it easy for readers to assess amplitudes and latencies. The voltage scale should indicate whether positive or negative is up (and if negative is up, I would recommend stating this in the figure caption). In addition, the electrode sites should be labeled in the figure (if a single site is shown, I often indicate the site in the caption).

Figure 6.1 shows an example of a bad way (panel A) and a good way (panel B) to plot ERP waveforms. The most egregious error in

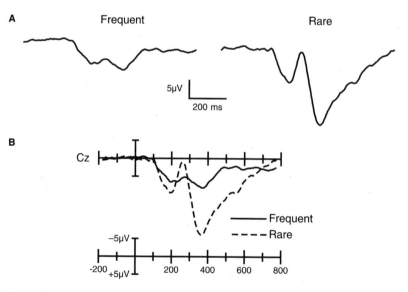

Figure 6.1   Examples of a poor way to plot ERP waveforms (A) and a good way to plot them (B). Negative is plotted upward.

panel A is that there is no X axis line running through zero micro-volts to provide an anchor point for the waveforms. It is absolutely essential to include an X axis line that shows the zero microvolt point and the time scale for each waveform. Without it, the reader will find it difficult to determine the amplitude or the latency of the various points in the waveform. A separate scale marker, such as that shown in figure 6.1A, is not enough, because it does not provide a visual reference point. For example, can you tell how much bigger the first component is in the left waveform than in the right waveform? And can you see that this component peaks earlier in the right waveform than in the left waveform?

The second most egregious error in figure 6.1A is that the waveforms are not overlapped. This makes it very difficult for the reader to determine the exact pattern of effects. For example, can you tell exactly when the difference between the waveforms switches polarities? All key effects should be shown by overlapping the relevant waveforms. Figure 6.1A also fails to indicate time zero, and it is impossible to tell which polarity is plotted upward.

Panel B of figure 6.1 solves these problems. It has an X axis running through the waveforms, which are overlapped. It provides additional calibration axis, indicating the latencies (with minor tick marks as well, showing the intermediate latencies). The voltage calibration indicates the polarity of the waveform. The two waveforms are drawn in easy-to-distinguish line types, and the figure provides a legend to indicate what condition each line type represents (this is much better than providing this information in words in the figure caption or main text). The figure also indicates the electrode site at which the waveforms were measured.

It is also essential for the plot to show a significant prestimulus portion of the waveforms. The prestimulus period allows the reader to assess the overall noise level of the recordings and the presence of overlap from the preceding trials. In addition, the apparent amplitude of the ERP peaks will depend on the baseline level, and a stable baseline is therefore essential for accurately characterizing the differences between two or more waveforms.

I almost always provide a 200-ms prestimulus baseline, and I would recommend this for the vast majority of cognitive ERP experiments.

You should be careful when selecting the line types for the ERP waveforms. First, you need to use reasonably thick lines. Figures are usually reduced by a factor of two or three times for publication, making lines that were originally moderately thin become so thin that they are nearly invisible. If someone makes a photocopy of the journal article, portions of the lines may disappear. Of course, you don't want to make the lines so thick that they obscure the experimental effects. I find that a two-point width works well in the majority of cases. In addition, when multiple waveforms overlap, you should make sure that the different lines are easy to differentiate. The best way to do this is with solid, dashed, and dotted lines (or with color, if available, but then photocopies will lose the information). You should avoid using different line thicknesses to differentiate the waveforms; this doesn't work as well, and the thin lines may disappear once the figure has been reduced for publication. In general, you should avoid overlapping more than three waveforms (four if absolutely necessary). With more overlapping waveforms, the figure starts to look like spaghetti.

How many different electrode sites should you show? This depends, in part, on the expected audience. If the paper will be published in an ERP-oriented journal, such as *Psychophysiology*, the readers will expect to see waveforms from a broad selection of electrode sites. If the paper will be published in a journal where ERP studies are not terribly common, such as Journal *of Neuroscience* or *JEP: General*, you should show only the key sites. If you have recorded from a large number of electrodes, it is pointless to include a figure showing the data from all the sites. If a figure contains more than ten to fifteen sites, the individual waveforms will be so small that they will provide very little information. It is much better to show a representative sample of the electrode sites and then use topographic maps to show the distribution of voltage over the scalp.

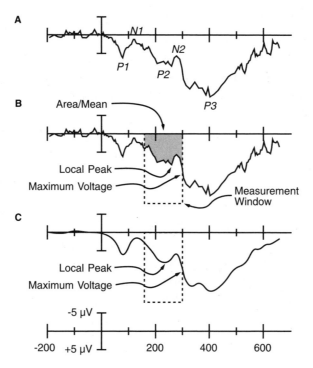

Figure 6.2   (A) ERP waveform with several peaks. (B) ERP waveform, with a measurement window from 150–300 ms poststimulus and several measures. (C) Filtered ERP waveform. Negative is plotted upward.

Measuring ERP Amplitudes

Most ERP studies concentrate on the amplitudes of one or more ERP components, and it is important to measure amplitude accurately. As figure 6.2 illustrates, there are two common ways to measure ERP amplitudes. The most common method is to define a time window and, for each waveform being measured, find the maximum amplitude in that time window. This is called a *peak amplitude* measure. The second most common method is to define a time window and, for each waveform being measured, calculate the mean voltage in that time window. This is called a *mean amplitude* measure. It is also possible to calculate the sum of the voltages at each time point within the measurement window (an *area amplitude* measure), but this is really just the mean amplitude multiplied by the number of points in the measurement window. For all of these measures, the voltages are typically measured relative to the average voltage in the prestimulus period. Other techniques are sometimes used, but they are beyond the scope of this book.

The goal of these techniques is to provide an accurate measurement of the size of the underlying ERP component with minimal distortion from noise and from other, overlapping components. As I discussed at the beginning of chapter 2, it is very difficult to measure the amplitude and latency directly from a raw ERP waveform without distortion from overlapping components. That chapter also described several strategies for avoiding this problem (e.g., using difference waves), but those strategies often rely on the use of an appropriate measurement approach. And don't forget rule 1 from chapter 2: *Peaks and components are not the same thing. There is nothing special about the point at which the voltage reaches a local maximum.*

Peak Amplitude

Given that the point at which the voltage reaches a local maximum is not special, why should you use peak amplitude to measure the amplitude of an ERP component? In fact, there is no compelling reason to use peak amplitude measures in most ERP experiments (although there might be good reasons to use this measure in some special cases). Moreover, there are several reasons why you shouldn't use peak amplitude in many cases, at least not without some modifications to the basic procedure.

Consider, for example, the ERP waveform shown in panel A of figure 6.2. To measure the peak of the P2 wave, you might use a time window of 150–300 ms. As panel B of figure 6.2 shows, the maximum voltage in this time window occurs at the edge of the time window (300 ms) due to the onset of the P3 wave. Consequently, the amplitude of the P2 wave would be measured at 300 ms, which is not very near the actual P2 peak. Clearly, this is not a good way to measure the P2 wave. You could avoid this problem by using a narrower measurement window, but a fairly wide window is usually necessary because of variations in peak latency across electrode sites, experimental conditions, and subjects. A much better way to avoid this problem is to search the measurement window for the largest point that is surrounded on both sides by smaller points. I call this measure *local peak amplitude*, and I refer to the original method as *simple peak amplitude*. To minimize spurious local peaks caused by high-frequency noise, the local peak should be defined as having a greater voltage than the average of the three to five points on either side rather than the one point on either side.[1]

Local peak amplitude is obviously a better measure than the simple peak amplitude. Strangely, I have never seen anyone explicitly describe this way of measuring peaks. It may be that people use it and just call it peak amplitude (I have done this myself), but it would probably be best if everyone used the term *local peak amplitude* to make it clear exactly how the peaks were measured.

Although local peak amplitude is a better measure than the simple peak amplitude, high-frequency noise can significantly distort both of these measures. In figure 6.2B, for example, the local peak is not in the center of the P2 wave, but is shifted to the right because of a noise deflection. It makes sense that peak amplitude would tend to be noise-prone, because the voltage at a single time point is used to measure a component that lasts for hundreds of milliseconds. Moreover, the peak amplitude measurements will tend to be larger for noisier data and for wider measurement windows, because both of these factors increase the likelihood that a really large noise deflection will occur by chance. Consequently, it is never valid to compare peak amplitudes from averages of different numbers of trials or from time windows of different lengths. I have seen this rule violated in several ERP papers, and it definitely biases the results.

If, for some reason, you still want to measure peak amplitude, you can reduce the effects of high-frequency noise by filtering out the high frequencies before measuring the peak. Panel C of figure 6.2, which shows the waveform in panels B and C after low-pass filtering, illustrates this. In the filtered waveform, the simple peak amplitude again provides a distorted measure, but the local peak provides a reasonably good measure of the amplitude. As discussed in chapter 5, when high frequencies are filtered out, the voltage at each point in the filtered waveform reflects a weighted contribution of the voltages from the nearby time points. Thus, the peak amplitude in a filtered waveform avoids the problem of using the voltage at a single time point to represent a component that lasts hundreds of milliseconds.

Another shortcoming of peak amplitude measures is that they are essentially nonlinear. For example, if you measure the peak amplitude of the P2 wave for several subjects and compute the average of the peak amplitudes, the result will almost always be different from what would you would get by creating a grand average of the single-subject waveforms and measuring the peak amplitude of the P2 wave in this grand average. This may cause a discrepancy between the grand-average waveforms that you present in your figures and the averaged peak amplitude values that you analyze statistically. Similarly, when there is variability in the latency of a component from trial to trial in the raw EEG data, the peak amplitude in the averaged ERP waveform will be smaller than the single-trial amplitudes (see panels G and H of figure 2.1 in chapter 2). If latency variability is greater in one condition than in another, the peak amplitudes will differ between conditions even if there is no difference between conditions in the single-trial peak amplitudes.

To summarize, peak amplitude measures have four serious shortcomings: (1) when the maximum voltage in the measurement window is measured, the rising or falling edge of an overlapping component at the border of the measurement window may be measured rather than the desired peak; (2) peak amplitude uses a single point to represent a component that may last hundreds of milliseconds, making it sensitive to noise; (3) peak amplitude will be artificially increased if the noise level is higher (due, for example, to a smaller number of trials) or if the measurement interval is longer; (4) peak amplitude is a nonlinear measure that may not correspond well with grand-average ERP waveforms or with single-trial peaks.

It might be tempting to argue that peak amplitude is a useful measure because it takes into account variations in latency among conditions, electrode sites, and subjects. After all, you wouldn't want to measure the amplitude of a component at the same time point at all electrode sites if the component peaks later at some sites than at others, would you? This argument is fallacious and points to a more abstract reason for avoiding peak amplitude measures. Specifically, peak amplitude measures implicitly encourage the mistaken view that peaks and components are the same thing and that there is something special about the point at which the voltage reaches its maximum value.

For example, the fact that the voltage reaches its maximum value at different times for different electrode sites is unlikely to reflect differences in the latency of the underlying component at different electrode sites. A component, as typically defined, is a single brain process that influences the recorded voltage simultaneously at all electrode sites, and it cannot have a different latency at different sites. As described in chapter 2 (see especially figure 2.1), the timing of a peak in the observed ERP waveform will vary as a function of the relative amplitudes of the various overlapping components, and this is the reason why peak latencies vary across electrode sites. Indeed, it would be rather strange to measure the same component at different times for the different electrode sites. Similarly, peak latency differences between subjects are just as likely to be due to differences in the relative amplitudes of the various underlying components as differences in the latency of the component of interest. In contrast, the latency of the underlying component may indeed vary considerably across different experimental conditions. But in this case, it will be nearly impossible to measure the component without significant distortion from other overlapping components, because amplitude and latency measures are often confounded (see chapter 2). Thus, although peak amplitude measures might seem less influenced by changes in component timing, this is really an illusory advantage.

Mean Amplitude

Mean amplitude has several advantages over peak amplitude. First, you can use a narrower measurement window because it doesn't matter if the maximum amplitude falls outside of this window for some electrode sites or some subjects. In fact, the narrower the window, the less influence overlapping components will have on the measurements. As an example, a measurement window of 200–250 ms would be appropriate for measuring the mean amplitude of the P2 component in the waveform shown in figure 6.2, compared to a window of 150–300 ms that would be appropriate for a peak amplitude measure.

A second advantage of mean amplitude measures is that they tend to be less sensitive to high-frequency noise than are peak amplitude measures, because a range of time points is used rather than a single time point. For this reason, you don't want to make your measurement windows too narrow (< 40 ms or so), even though a narrow window is useful for mitigating the effects of overlapping components. You should also note that, although filtering can reduce the effects of high-frequency noise in peak amplitude measures, there is no advantage to low-pass filtering when measuring mean amplitude. By definition, mean amplitude includes the voltages from multiple nearby time points, and this is exactly what low-pass filtering does. In fact, filtering the data before measuring the mean amplitude in a given measurement window is the same thing as measuring the mean amplitude from an unfiltered waveform using a wider measurement window.

A third advantage of mean amplitude measures is that they do not become biased when the noise level increases or when one uses a longer measurement window. In other words, the variance may change, but the expected value is independent of these factors. Consequently, it is legitimate to compare mean amplitude measurements from waveforms based on different numbers of trials, whereas this is not legitimate for peak amplitude measurements.

A fourth advantage is that mean amplitude is a linear measure. That is, if you measure the mean amplitude of a component from each subject, the mean of these measures will be equal to measuring the mean amplitude of the component from the grand-average waveform. This makes it possible for you to compare directly your grand-average waveforms with the means from your statistical analyses. This same principle also applies to the process of averaging together the single-trial EEG data to form averaged ERP waveforms. That is, the mean amplitude measured from a subject's averaged ERP waveform will be the same as the average of the mean amplitudes measured from the single-trial EEG data.

Although mean amplitude has several advantages over peak amplitude, it is not a panacea. In particular, mean amplitude is still quite sensitive to the problem of overlapping components and can lead to spurious results if the latency of a component varies across conditions. In addition, there is not always an a priori reason to select a particular measurement window, and this can encourage fishing for significant results by trying different windows. There are some more sophisticated ways of measuring the amplitude of a component (e.g., dipole source modeling, ICA, PCA, etc.), but these methods are based on a variety of difficult-to-assess assumptions and are beyond the scope of this book (for more information on alternative measures, see Coles et al., 1986). Thus, I would generally recommend using mean amplitude measures in conjunction with the rules and strategies for avoiding the problem of overlapping components discussed in chapter 2.

Baselines

Whether you are measuring mean amplitude or peak amplitude, you will be implicitly or explicitly subtracting a voltage—usually the average prestimulus voltage—that represents the baseline or zero point. Any noise in the baseline will therefore add noise to your measures, so it is important to select an appropriate baseline. When you are measuring stimulus-locked averages, I would recommend that you use the average voltage in the 200 ms before stimulus onset as the baseline. A 100-ms baseline is common, and although it's not as good as 200 ms, it's usually acceptable. If you use less than 100 ms, it is likely that you will be adding noise to your measures.

The prestimulus interval is usually used as the baseline because it is assumed that the voltage in this period is unaffected by the stimulus. Although it is true that the time-locking stimulus cannot influence the prestimulus period, it is important to realize that the processing does not always begin after the onset of a stimulus. If the interval between stimuli is relatively short (< 2 s), the late potentials from the previous stimulus may not have completely diminished by the time of the time-locking stimulus and may therefore contribute to the baseline voltage. In addition, regardless of the length of the interstimulus interval, the prestimulus voltage may be influenced by preparatory processes (in fact, the effects of preparatory processes are more visible when the interstimulus interval is long). For these reasons, the prestimulus baseline rarely provides a perfectly neutral baseline, and in many experiments it is clear that the voltage slopes upward or downward during the prestimulus interval because of these factors. This can be a major problem if the prestimulus activity differs across experimental conditions, because any difference in measured amplitudes between conditions might reflect prestimulus differences rather than post-stimulus differences (see chapter 4, and Woldorff, 1988).

Even when the prestimulus activity does not differ across conditions, it is important to recognize that the prestimulus interval is usually not completely neutral. For example, scalp distributions can be significantly distorted if significant activity is present in the prestimulus interval and then fades by the time of the component being measured. The only real solution is to keep in mind that the measured voltage reflects the difference between the amplitude in the measurement window and the amplitude in the prestimulus period (just as it also reflects the difference between the active and reference electrodes).

Baselines become much more complicated when using response-locked averages, because the activity that precedes the response is often as large as or larger than the activity that follows the response. One solution to this problem is to use as the baseline a period of time that precedes the response by enough so that it always precedes the stimulus. Another solution is to use the average voltage of the entire averaging epoch as the baseline. As far as I am aware, there is no single best solution, other than thinking carefully about whatever baseline period you use to make sure that it isn't distorting your results.

I should also mention that investigators sometimes compute *peak-to-peak* amplitudes that consist of the difference in amplitude between two successive peaks of opposite polarity. For example, the amplitude of the visual N1 wave is sometimes measured as the difference in amplitude between the N1 peak and the preceding P1 peak (this could also be done with mean amplitudes, although this is much less common). This approach is sometimes taken if the components are assumed to overlap in time such that the earlier component distorts the amplitude of the later component. This may be a reasonable approach, as long as you keep in mind exactly what you are measuring.

Measuring ERP Latencies

Peak Latency

ERP latencies are usually measured with *peak latency* measures that find the maximum amplitude within a time window and use the latency of this peak as a measure of the latency of the underlying component. All of the shortcomings of peak amplitude measures also have analogs in peak latency measures. First, when the measurement window includes the rising or falling edge of a larger component, the maximum voltage will be at the border of the window (see figure 6.2B). Fortunately, this problem can be solved by using a *local peak latency* measure in which a point is not considered a peak unless the three to five points on each side of it have smaller values.

A second shortcoming of peak latency measures is that, like peak amplitude measures, they are highly sensitive to high-frequency noise. If a peak is rather broad and flat, high-frequency noise may cause the maximum voltage to be very far away from the middle of the peak (see the local peak in figure 6.2B). In fact, this is probably an even more significant problem for peak latency than for peak amplitude, because a noise-related peak may be far in time from the true peak but close to the true peak's amplitude. As in the case of peak amplitude, you can mitigate this problem to some extent by filtering out the high frequencies in the waveform (see the local peak in figure 6.2C).

A third shortcoming is that peak latency, like peak amplitude, will change systematically as noise levels increase. Specifically, as the noise increases, the average peak latency will tend to be nearer to the center of the measurement window. To understand why this is true, imagine an ERP waveform that is entirely composed of random noise. If you measure the peak latency between 200 and 400 ms, the peak latency on any given trial is equally likely to be at any value in this range, and the average will therefore be at the center of the range. If there is a signal as well as noise, then the average will tend to be somewhere between the actual peak and the center of the range.

A fourth shortcoming of peak latency is that it is nonlinear. In other words, the peak latency measured from a grand average will not usually be the same as the average of the peak latencies that were measured from the single-subject waveforms. Similarly, the peak latency measured from a single-subject average will not usually be the same as the average of the single-trial peak latencies.

A fifth shortcoming is that peak latency measures, like peak amplitude measures, implicitly encourage the mistaken view that peaks and components are the same thing and that there is something special about the point at which the voltage reaches its maximum value. The latency at which the voltage reaches its maximum value depends greatly on the nature of the overlapping components and the waveshape of the component of interest, so peak latency bears no special relationship to the timing of an ERP component.

Although peak latency has many shortcomings, there just aren't many good alternatives, and so it is often the best measure. When measuring peak latency, you should take the following precautions: (1) filter out the high-frequency noise in the waveforms; (2) use a local peak measure rather than an absolute peak measure; (3) make sure that the waveforms being compared have similar noise levels; and (4) keep in mind that peak latency is a coarse and nonlinear measure of a component's timing.

Fractional Area Latency

Under some conditions, it is possible to avoid some of the shortcomings of peak latency measures by using *fractional area latency* measures, which are analogous to mean amplitude measures. Fractional area measures work by computing the area under the ERP waveform over a given latency range and then finding the time point that divides that area into a prespecified fraction (this approach was apparently first used by Hansen & Hillyard, 1980). Typically the fraction will be a half, in which case this would be called a *50 percent area latency* measure. Figure 6.3A shows an example of this. The measurement window in this figure is 300–600 ms, and the area under the curve in this time window is divided at 432 ms into two regions of equal area. Thus, the 50 percent area latency is 432 ms.

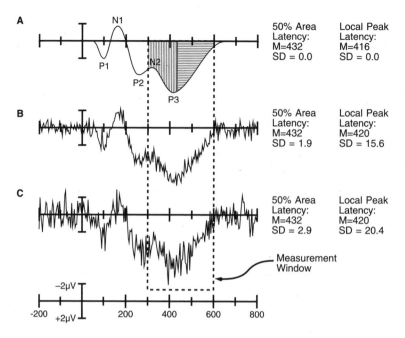

Figure 6.3 Application of 50 percent area latency and local peak latency measures to a noise-free ERP waveform (A), an ERP waveform with a moderate amount of noise (B), and an ERP waveform with significant noise (C). In each case, measurements were obtained from 100 waveforms, and the mean (M) and standard deviation (SD) across these 100 measures are shown. Negative is plotted upward.

The latency value estimated in this manner will depend quite a bit on the measurement window chosen. For example, if the measurement window for the waveform shown in figure 6.3A was shortened to 300–500 ms rather than 300–600 ms, an earlier 50 percent area latency value would have been computed. Consequently, this measure is not appropriate for estimating the absolute latency of a component unless the measurement window includes the entire component and no other overlapping components are present. However, peak latency also provides a poor measure of absolute latency because it is highly distorted by overlapping components and is relatively insensitive to changes in waveshape. Fortunately, in the vast majority of experiments, you don't really care about the absolute latency of a component and are instead concerned with the relative latencies in two different conditions. Consequently, when you are deciding how to measure the latencies in an experiment, you should base your decision on the ability of a measure to accurately characterize the difference in latency between two conditions. In many cases, the 50 percent area latency measure can accurately quantify latency differences even when the measurement window does not contain the entire component and when there is some overlap from other components.

One advantage of the 50 percent area latency measure is that it is less sensitive to noise. To demonstrate this, I added random (Gaussian) noise the waveform shown in figure 6.3B and then measured the 50 percent area latency and the local peak latency of the P3 wave. I did this a hundred times for each of two noise levels, making it possible to estimate the variability of the measures. When the noise level was 0.5 µV, the standard deviation of the peak latency measure over the hundred measurements was 15.6 ms, whereas the standard deviation of the 50 percent area latency measure was only 1.9 ms. This is shown in figure 6.3B, which shows the waveform from one trial of the simulation (the basic waveform was the same on each of the hundred trials, but the random noise differed from trial to trial). When the noise level was increased to 1.0 µV (figure 6.3C), the standard deviation of the peak latency measure was 20.4 ms, whereas the standard deviation of the 50 percent area latency measure was only 2.9 ms. The variability in peak latency measures can be greatly decreased by filtering the data, and a fair test of peak latency should be done with filtered data. When the waveforms were filtered by convolving them with a Gaussian impulse response function (SD = 30 ms, half-amplitude cutoff at 6 Hz), the standard deviations of the peak latency measures dropped to 3.3 ms and 6.1 ms for noise levels of 0.5 and 1.0 µV, respectively. Thus, even when the data were filtered, the standard deviation of the peak latency measure was approximately twice as large as the standard deviation of the 50 percent area latency measure.

The 50 percent area latency measure has several other advantages as well. First, it is a sensible way to measure the timing of a component that doesn't have a distinct peak or has multiple peaks. Second, it has the same expected value irrespective of the noise level of the data. Third, the 50 percent area latency is linear, so the mean of your measures will correspond well with what you see in the grand-average waveforms. Finally, 50 percent area latency can be related to reaction time more directly than peak latency, as described in the next section.

Although the 50 percent area latency measure has several advantages over peak latency, it also has a significant disadvantage. Specifically, it can produce very distorted results when the latency range does not encompass most of the ERP component of interest or when the latency range includes large contributions from multiple ERP components. Unfortunately, this precludes the use of the 50 percent area latency measure in a large proportion of experiments. It is useful primarily for late, large components such as P3 and N400, and primarily when the component of interest can be isolated by means of a difference wave. In my own research, for example, I have used the 50 percent area latency measure in only two studies (Luck, 1998b; Luck & Hillyard, 1990). However, these were the two studies in which I was most interested in measuring ERP latencies and comparing the results with RT. Thus, 50 percent area latency can be a very useful measure, but it is somewhat limited and works best in experiments that have been optimized to isolate a specific ERP component.

I would like to end this section by emphasizing that it is very difficult to measure ERP latencies. The peak latency measure has several problems, and although the 50 percent area latency measure has some advantages over peak latency, it is appropriate only under a restricted set of conditions. Moreover, the sophisticated PCA and ICA techniques that have been developed for measuring ERP amplitudes cannot be used to measure latencies (and generally assume that latencies are constant). Thus, you must be extremely careful when interpreting latency measures.

Comparing ERP Latencies with Reaction Times

In many studies, it is useful to compare the size of an ERP latency effect to the size of a reaction time (RT) effect. This seems straightforward, but it is actually quite difficult. The problem is that RT is usually summarized as the mean across many trials, whereas a peak in an ERP waveform is more closely related to the peak (i.e., the mode) of the distribution of single-trial latencies, and 50 percent area latency is more closely related to the median of this distribution. In many experiments, RT differences are driven by changes in the tail of the RT distribution, which influences the mean much more than the mode or the median. Consequently, ERP latency effects are commonly smaller than mean RT effects.

To make this clearer, figure 6.4A shows the probability distribution of RT in two conditions of a hypothetical experiment, which we'll call the *easy* and *difficult* conditions. Each point represents the probability of an RT occurring with ± 15 ms of that time (that is, the figure is a histogram with a bin width of 30 ms). As is typical, the RT distributions are skewed, with a right-side tail extending out to long RTs, and much of the RT difference between the conditions is due to a change in the probability of relatively long RTs rather than a pure shift in the RT distribution. Imagine that the P3 wave in this experiment is precisely time-locked to the response, always peaking 150 ms after the RT. Consequently, the P3 wave will occur at different times on different trials, with a probability distribution that is shaped just like the RT distribution from the same condition (but shifted rightward by 150 ms). Imagine further that the earlier components are time-locked to the stimulus rather than the response. Figure 6.4B shows the resulting averaged ERP waveforms for these two conditions.

Because most of the RTs occur within a fairly narrow time range in the easy condition, most of the single-trial P3s will also occur within a narrow range, causing the peak of the averaged ERP waveform to occur approximately 150 ms after the peak of the RT distribution (overlap from the other components will influence the precise latency of the peak). Some of the single-trial RTs occur

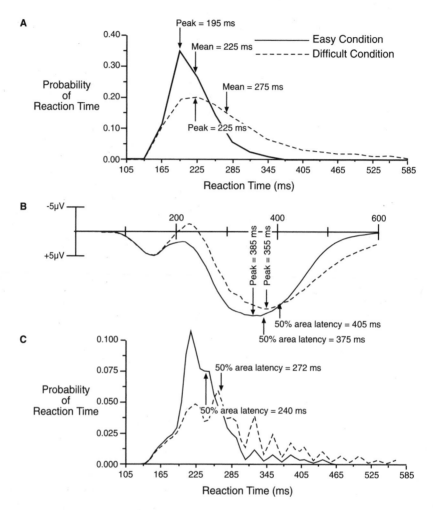

Figure 6.4   (A) Histogram showing the probability of a response occurring in various time bins (bin width = 30 ms) in an easy condition and a difficult condition of a hypothetical experiment. (B) ERP waveforms that would be produced in this hypothetical experiment if the early components were insensitive to reaction time and the P3 wave was perfectly time-locked to the responses. Negative is plotted upward. (C) Response density waveforms showing the probability of a response occurring at any given moment in time.

at longer latencies, but they are sufficiently infrequent that they don't have much influence on the peak P3 latency in the averaged waveform.

The mean RT is 50 ms later in the difficult condition than in the easy condition. However, because much of the RT effect consists of an increase in long RTs, the peak of the RT distribution is only 30 ms later in condition B than in condition A. Because the peak of the P3 wave in the averaged ERP waveform is tied closely to the peak of the RT distribution, peak P3 latency is also 30 ms later in the difficult condition than in the easy condition. Thus, the peak latency of the P3 wave changes in a manner that reflects changes in the peak (mode) of the RT distribution rather than its mean. Thus, when RT effects consist largely of increases in the tail of the distribution rather than a shift of the whole distribution, changes in peak latency will usually be smaller than changes in mean RT, even if the component and the response are influenced by the experimental manipulation in exactly the same way. Consequently, you should never compare the magnitude of an ERP peak latency effect to a mean RT effect unless you know that the RT effect is simply a rightward shift in the RT distribution (which you can verify by plotting the distributions, as in figure 5.4A).

How, then, can you compare RT effects to ERP latency effects? The answer is you must measure them in the same way. One way to achieve this would be to use a peak latency measure for both the ERPs and the RTs (using the probability distribution to find the peak RT). However, the peak of the RT distribution is unlikely to reflect the RT effects very well, because the effects are often driven by changes in the longer RTs. In addition, RT distributions are typically very noisy unless they are computed from thousands of RTs, so it is difficult to get a stable measure of the peak of the RT distribution in most experiments. Moreover, the latency of a peak measured from an ERP waveform is not a particularly good measure of the timing of the underlying component, as discussed earlier in this chapter.

The 50 percent area latency measure provides the time point that bisects the area of the waveform, and this is similar to the median RT, which is the point separating the fastest half of the RTs from the slowest half. Indeed, I once conducted a study which compared the 50 percent area latency of the P3 wave with median RT, and the results were quite good (Luck, 1998b). However, 50 percent area latency is not quite analogous to median RT, because median RT does not take into account the precise values of the RTs above and below the median. For example, a median RT of 300 ms would be obtained for an RT distribution in which half of the values were between 200 and 300 ms and the other half were between 300 and 400 ms, and the same median RT would be obtained if half of the RTs were between 290 and 300 ms and the other half were between 300 and 5,000 ms.

Thus, we need a measure of RT that represents the point that bisects the *area* of the RT distribution, just as the 50 percent area latency measure represents the point that bisects the area under the ERP waveform. The problem is that RTs are discrete, instantaneous events, and this is a problem for area measures. One way to achieve a curve with a measurable area would be to compute histograms of RT probability, as in figure 6.4A. However, these histograms are not really continuous, so they do not provide a perfect measure of area. Researchers who conduct single-unit recordings have to deal with this same problem, because spikes are also treated as discrete, instantaneous events. To create a continuous waveform from a set of spikes, they have developed *spike density waveforms* (see, e.g. Szücs, 1998). In these waveforms, each spike is replaced by a continuous gaussian function, peaking at the time of the spike, and the gaussians are summed together. The same thing can be done for reaction times, creating *response density waveforms*.

Figure 6.4C shows an example of such a waveform. Each individual response that contributed to the histogram in figure 6.4A was replaced by a Gaussian function with a standard deviation of approximately 8 ms, and the average across trials was then computed. The 50 percent area latency was then computed, just as it was for the ERPs. The resulting latencies were 240 ms for the easy condition and 272 ms for the difficult condition, and the 32-ms difference between these latencies compares favorably with the 30-ms effect observed in the 50 percent area latency measures of the ERP waveform.

I don't know of anyone who has tried this approach, and I'm sure it has limitations (I described some of the limitations of the 50 percent area latency measure toward the end of the previous section). However, this approach to comparing RT effects with ERP latency effects is certainly better than comparing peak ERP latencies with mean RTs.

Onset Latency

In many experiments, it is useful to know if the onset time of an ERP component varies across conditions. Unfortunately, onset latency is quite difficult to measure. When the onset of the component of interest is overlapped by other ERP components, there is no way to accurately measure the onset time directly from the waveforms. To measure the onset time, it is necessary to somehow remove the overlapping components. The most common way to do this is to form difference waves that isolate the component of interest (such as the difference waves used to extract the lateralized readiness potential, as described in chapter 2). More sophisticated approaches, such as dipole source analysis and independent components analysis may also be used to isolate an ERP component.

Once a component has been isolated, it is still quite difficult to measure its onset time. The main reason for this is that the onset is a point at which the component's amplitude is, by definition, at or near zero, and the signal-to-noise ratio is also at or near zero. Thus, any noise in the waveform will obscure the actual onset time. Researchers have used several different methods to measure onset time over the years; I'll describe a few of the most common methods.

One simple approach is to plot the waveforms for each subject on paper and have a naïve individual use a straight-edge to extrapolate the waveform to zero μV and then record the latency at that point. This can be time consuming, and it assumes that the component can be accurately characterized as a linearly increasing waveform. Despite the fact that this approach is somewhat subjective, it can work quite well because it takes advantage of the sophisticated heuristics the human visual system uses. But it is also prone to bias, so the person who determines the latencies must be blind to the experimental conditions for each waveform.

Another approach is to use the fractional area latency measure described previously in this chapter. Instead of searching for the point that divides the waveform into two regions of equal area, it is possible to search for the point at which 25 percent of the area has occurred (see, e.g., Hansen & Hillyard, 1980). This can work well, but it can be influenced by portions of the component that occur long after the onset of the component. Imagine that waveforms A and B have the same onset time, but waveform B grows larger than waveform A at a long latency. Waveform B will therefore have a greater overall area than waveform A, and it will take longer to reach 25 percent of the area for waveform B.

Another approach is to find the time at which the waveform's amplitude exceeds the value expected by chance. The variation in prestimulus voltage can be used to assess the amplitude required to exceed chance, and the latency for a given waveform is the time at which this amplitude is first reached (for details, see Miller, Patterson, & Ulrich, 1998; Osman et al., 1992). Unfortunately, this method is highly dependent on the noise level, which may vary considerably across subjects and conditions.

A related approach is to find the time at which two conditions become significantly different from each other across subjects, which provides the onset latency of the experimental effect. The simplest way to achieve this is to perform a t-test of the two conditions at each time point and find the first point at which the corresponding p-value is less than .05. The problem with this is that many comparisons are being made, making it likely that a spurious value will be found. One could use an adjustment for multiple comparisons, but this usually leads to a highly conservative test that will become significant much later than the actual onset time of the effect. A compromise approach is to find the first time point that meets two criteria: (1) the p-value is less than .05, and (2) the p-values for the subsequent N points are also less than .05 (where N is usually in the range of 3–10). The idea is that it is unlikely to have several spurious values in a row, so this approach won't pick up spurious values. That is fairly reasonable, but it assumes that the noise at one time point is independent of the noise at adjacent time points. This is generally not true, and it is definitely false if the data have been low-pass filtered. As discussed in chapter 5, low-pass filtering spreads the voltage out in time, and a noise blip at one point in time will therefore be spread to nearby time points. However, if the data have not been extensively low-pass filtered, and N is sufficiently high (e.g., 100 ms), this approach will probably work well.

Miller, Patterson, and Ulrich (1998) have developed an excellent approach for measuring the difference in the onset latency of the lateralized readiness potential between two conditions, and this approach should also be useful for other components that can be isolated from the rest of the ERP waveform. In this approach, one uses the grand-average waveforms from two conditions and finds the point at which the voltage reaches a particular level (e.g., 50 percent of the peak amplitude of the waveform). The difference between the two conditions provides an estimate of the difference in onset times. Because this estimate is based on the point at which a reasonably large amplitude has been reached in the grand average, it is relatively insensitive to noise. One can then test the statistical significance of this estimate by using the *jackknife technique* to assess the standard error of the estimate (see Miller, Patterson, & Ulrich, 1998 for details). This is a well-reasoned technique, and Miller and colleagues (1998) performed a series of simulations to demonstrate that it works well under realistic conditions.

Statistical Analyses

Once you have collected ERP waveforms from a sample of subjects and obtained amplitude and latency measures, it is time to perform some statistical analyses to see whether your effects are significant. In the large majority of cognitive ERP experiments, the investigators are looking for a main effect or an interaction in a completely crossed factorial design, and ANOVA is therefore the dominant statistical approach. Consequently, this is the only approach I will describe. Other approaches are sometimes useful, but they are beyond the scope of this book.

Before I begin describing how to use ANOVAs to analyze ERP data, I would like to make it clear that I consider statistics to be a necessary evil. We often treat the .05 alpha level as being somehow magical, with experimental effects that fall below $p < .05$ as being "real" and effects that fall above $p < .05$ as being nonexistent. This is, of course, quite ridiculous. After all, if $p = .06$, there is only a 6 percent probability that the effect was due to chance, which is still rather low. Moreover, the assumptions of ANOVA are violated by almost every ERP experiment, so the p-values that we get are only approximations of the actual probability of a type I error. On the other hand, when examining a set of experimental effects that are somewhat weak, you need a criterion for deciding whether to believe them or not, and a p-value is usually better than nothing.

I have two specific recommendations for avoiding the problems associated with conventional statistics. First, whenever possible, try to design your experiments so that the experimental effects are quite large relative to the noise level and the p-values are very low (.01 or better). That way, you won't have to worry about one "bad" subject keeping your p-values from reaching the magical .05 criterion. Moreover, when your effects are large relative to the noise level, you can have some faith in the details of the results that are not being analyzed statistically (e.g., the onset and offset times of the experimental effects). My second suggestion is to use the one statistic that cannot fail, namely replication (see box 6.1). If you

Box 6.1   The Best Statistic

Replication is the best statistic. I learned this when I was a graduate student in the Hillyard lab, although no one ever said it aloud. I frequently say it aloud to my students. Replication does not depend on assumptions about normality, sphericity, or independence. Replication is not distorted by outliers. Replication is a cornerstone of science. Replication is the best statistic.

A corollary principle—which Steve Hillyard *has* said aloud—is that the more important a result is, the more important it is to replicate the result before you publish it. There are two reasons for this, the first of which is obvious: you don't want to make a fool of yourself by making a bold new claim and being wrong. The second and less obvious reason is that if you want people to give this important new result the attention it deserves, you should make sure that they have no reason to doubt it. Of course, it's rarely worthwhile to run exactly the same experiment twice. But it's often a good idea to run a follow-up experiment that replicates the result of the first experiment and also extends it (e.g., by assessing its generality or ruling out an alternative explanation).

keep getting the same effect in experiment after experiment, then it's a real effect. If you get the effect in only about half of your experiments, then it's a weak effect and you should probably figure out a way to make it stronger so that you can study it more easily.

The Standard Approach

To explain the standard approach to analyzing ERP data with ANOVAs, I will describe the analyses that we conducted for the experiment described briefly near the beginning of chapter 1. In this experiment, we obtained recordings from lateral and midline electrode sites at frontal, central, and parietal locations (i.e., F3, Fz, F4, C3, Cz, C4, P3, Pz, and P4). Subjects saw a sequence of Xs and Os, pressing one button for Xs and another for Os. On some trial blocks, X occurred frequently (p = .75) and O occurred infrequently (p = .25), and on other blocks this was reversed. We also manipulated the difficulty of the X/O discrimination by varying the brightness of the stimuli. The experiment was focused on the P3 wave, but I will also discuss analyses of the P2 and N2 components.

Before measuring the amplitudes and latencies of these components, we first combined the data from the Xs and Os so that we had one waveform for the improbable stimuli and one for the probable stimuli. We did this for the simple reason that we didn't care if there were any differences between Xs and Os per se, and collapsing across them reduced the number of factors in the ANOVAs. The more factors are used in an ANOVA, the more individual p-values will be calculated, and the greater is the chance that one of them will be less than .05 due to chance (this is called an increase in *experimentwise* error). By collapsing the data across irrelevant factors, you can avoid this problem (and avoid having to come up with an explanation for a weird five-way interaction that is probably spurious).

Figure 6.5 illustrates the results of this experiment, showing the ERP waveforms recorded at Fz, Cz, and Pz. From this figure, it is clear that the P2, N2, and P3 waves were larger for the rare stimuli than for the frequent stimuli, especially when the stimuli were bright. Thus, for the amplitude of each component, we would expect to see a significant main effect of stimulus probability and a significant probability × brightness interaction.

The analysis of P3 amplitude is relatively straightforward. We measured P3 amplitude as the mean amplitude between 300 and 800 ms at each of the nine electrode sites and entered these data into a within-subjects ANOVA with four factors: stimulus probability (frequent vs. rare), stimulus brightness (bright vs. dim), anterior-posterior electrode position (frontal, central, or parietal), and left-right electrode position (left hemisphere, midline, or right hemisphere). We could have used a single factor for the electrode sites, with nine levels, but it is usually more informative to divide the electrodes into separate factors representing different spatial dimensions. Consistent with the waveforms shown in figure 6.5, this ANOVA yielded a highly significant main effect of stimulus

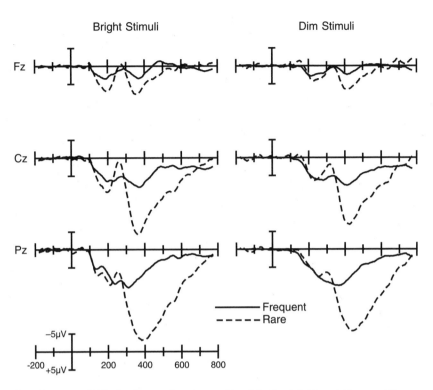

Figure 6.5 Grand-average ERPs for frequent and rare stimuli in the bright and dim conditions. Negative is plotted upward.

probability, $F(1, 9) = 95.48$, p < .001, and a significant interaction between probability and brightness, $F(1, 9) = 11.66$, p < .01.

Some investigators perform a separate ANOVA for each electrode site (or each left-midline-right set) rather than performing a single ANOVA with electrode site as a factor. Although there may be advantages to this approach, it is likely to increase both the probability of a type I error (incorrectly rejecting the null hypothesis) and the probability of a type II error (incorrectly accepting the null hypothesis). Type I errors will be increased because more p-values must be computed when a separate ANOVA is performed for each electrode, leading to a greater probability that a spurious effect will reach the .05 level. Type II errors will be increased because a small effect may fail to reach significance at any individual site, whereas the presence of this effect at multiple sites may be enough to make the effect significant when all of the sites contribute to the analysis.

Even when a single ANOVA includes multiple electrode sites, it is usually best not to include measurements from electrode sites spanning the entire scalp. Instead, it is usually best to measure and analyze an ERP component only at sites where the component is actually present. Otherwise, the sites where the component is absent may add noise to the analyses, or the presence of other components at those sites may distort the results. In addition, it is sometimes useful to analyze only the sites at which the component of interest is large *and* other components are relatively small so that they do not distort measurements of the component of interest. In the present study, for example, we used all nine sites for analyzing the P3 wave, which was much larger than the other components, but we restricted the P2 analyses to the frontal sites, where the P2 effects were large but the N2 and P3 waves were relatively small. However, when you are trying to draw conclusions about the scalp distribution of a component, it may be necessary to include measurements from all of the electrodes.

Box 6.2 provides a few thoughts about how to describe ANOVA results in a journal article.

Interactions with Electrode Site

It is clear from figure 6.5 that the difference in P3 amplitude between the rare and frequent stimuli was larger at posterior sites than at anterior sites. This led to a significant interaction between stimulus probability and anterior-posterior electrode position, $F(2, 18) = 63.92$, p < .001. In addition, the probability effect for the bright stimuli was somewhat larger than the probability effect for the dim stimuli at the parietal electrodes, but there wasn't much difference at the frontal electrodes. This led to a significant

Box 6.2 The Presentation of Statistics

Inferential statistics such as ANOVAs are used to tell us how much confidence we can have in our data. As such, they are not data, but are a means of determining whether the data pattern is believable. However, many ERP results sections are written as if the inferential statistics are the primary results, with virtually no mention of the descriptive statistics or the ERP waveforms. For example, consider this excerpt from a Results section:

Results
There was a significant main effect of probability on P3 amplitude,
$F(1, 9) = 4.57$, $p < .02$. *There was also a significant main effect of
electrode site,* $F(11, 99) = 3.84$, $p < .01$, *and a significant interaction
between probability and electrode site,* $F(11, 99) = 2.94$, $p < .05$ ...

In this example, the reader learns that there is a significant effect of probability on the P3 wave but cannot tell whether the P3 was bigger for the probable stimuli or for the improbable stimuli. This is not an unusual example: I have seen Results sections like this many times when reviewing journal submissions. It is important to remember that inferential statistics should always be used as support for the waveforms and the means, and the statistics should never be given without a description of the actual data. And I am not alone in this view. For example, an editorial in *Perception & Psychophysics* a few years ago stated that "the point of Results sections, including the statistical analyses included there, is to make the outcome of the experiment clear to the reader.... Readers should be directed first to the findings, then to their analysis" (Macmillan, 1999, p. 2).

three-way interaction among probability, brightness, and anterior-posterior electrode position, $F(2, 18) = 35.17$, p $< .001$.

From this interaction, you might be tempted to conclude that different neural generator sites were involved in the responses to the bright and dim stimuli. However, as McCarthy and Wood (1985) pointed out, ANOVA interactions involving an electrode position factor are ambiguous when two conditions have different mean amplitudes. Figure 6.6 illustrates this, showing the scalp distributions that one would expect from a single generator source in two different conditions, A and B. If the magnitude of the generator's

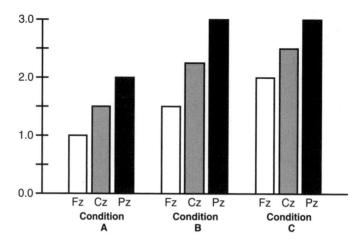

Figure 6.6   Examples of additive and multiplicative effects on ERP scalp distributions. Condition B is the same as condition A, except that the magnitude of the neural generator site has increased by 50 percent, thus increasing the voltage at each site by 50 percent. This is a multiplicative change. Condition C is the same as condition A, except that the voltage at each site has been increased by 1 μV. This is an additive change and is not what would typically be obtained by increasing the magnitude of the neural generator source.

activation is 50 percent larger in condition B than in condition A, the amplitude at each electrode will be 50 percent larger in condition B than in condition A. This is a multiplicative effect, and not an additive effect. That is, the voltage increases from 1 μV to 1.5 μV at Fz and increases from 2 μV to 3 μV at Pz, which is a 0.5 μV increase at Fz and a 1 μV increase at Pz. Condition C in figure 6.6 shows an additive effect. In this condition, the absolute voltage at each site increases by 1 μV, which is not the pattern that would result from a change in the amplitude of a single generator source. Thus, when a single generator source has a larger magnitude in one condition than in another condition, an interaction between condition and electrode site will be obtained (as in condition A versus condition B), whereas a change involving multiple generator sites may sometimes produce a purely additive effect (as in condition A versus condition C).

To determine whether an interaction between an experimental condition and electrode site really reflects a difference in the internal generator sources, McCarthy and Wood (1985) proposed *normalizing* the data to remove any differences in the overall amplitudes of the conditions. Specifically, the data are scaled so that the voltages across the electrode sites range between 0 and 1 in both conditions. To do this, you simply find the electrode sites with the largest and smallest amplitudes in each condition, and for each condition compute a new value according to the formula:

New value = (old value − minimum value)

÷ (maximum value − minimum value)

Once these normalized values have been computed, there will be no difference in amplitude between the conditions, and the condition × electrode site interaction is no longer distorted by the multiplicative relationship between the magnitude of the internal generator and the distribution of voltage across electrodes. Thus, any significant interaction obtained with the normalized values must be due to a change in the relative distribution of internal brain activity.

But there is a problem. Many labs used this procedure for almost two decades, and then Urbach and Kutas (2002) convincingly demonstrated that it doesn't work. For a variety of subtle technical reasons, normalization procedures will fail to adjust accurately for the multiplicative interactions that arise when ANOVAs are conducted with electrode as a factor. A significant condition × electrode site interaction may be obtained after normalization even if there is no difference in the relative distribution of internal brain activity, and a real difference in the relative distribution of internal brain activity may not yield a significant condition × electrode site interaction (even with infinite statistical power). Thus, you should not use normalization procedures, and there is simply no way to determine the presence or absence of a change in the relative distribution of internal brain activity by examining condition × electrode site interactions. It just cannot be done.

There is also another problem with normalization, but it's conceptual rather than technical. Many researchers have looked at condition × electrode site interactions to determine whether or not the same brain areas are active in different experimental conditions. Even if normalization worked correctly, it would be impossible to make claims of this sort on the basis of a condition × electrode site interaction. The problem is that this interaction will be significant when exactly the same generators are active in both conditions as long as they differ in relative magnitude. That is, if areas A and B have amplitudes of six and twelve units in one condition and eight and nine units in another condition, this will lead to a change in scalp distribution that will produce a condition × electrode site interaction. Moreover, even if there is no difference in the magnitude of the generators across conditions, but there is a latency difference in one of the two generators across conditions, it is likely that the measured scalp distribution at any given time point will differ across conditions. Thus, you cannot draw strong conclusions about differences in which generator sources are active on the basis of ANOVA results.

Violation of ANOVA Assumptions

It is well known that the ANOVA approach assumes that the data are normally distributed and that the variances of the different conditions are identical. These assumptions are often violated, but ANOVA is fairly robust when the violations are mild to moderate, with very little change in the actual probability of a type I error (Keppel, 1982). Unless the violations of these assumptions are fairly extreme (e.g., greater than a factor of two), you just don't need to worry about them. However, when using within-subjects ANOVAs, another assumption is necessary, namely homogeneity of covariance (also called *sphericity*). This assumption applies only when there are at least three levels of a factor. For example, imagine an experiment in which each subject participates in three conditions, C1, C2, and C3. In most cases, a subject who tends to have a high value in C1 will also tend to have high values in C2 and C3; in fact, this correlation between the conditions is the essential attribute of a within-subjects ANOVA. The assumption of homogeneity of covariance is simply the assumption that the degree of correlation between C1 and C2 is equal to the degree of correlation between C2 and C3 and between C1 and C3. This assumption does not apply if there are only two levels of a factor, because there is only one correlation to worry about in this case.

The homogeneity-of-covariance assumption is violated more often by ERP experiments than by most other types of experiments because data from nearby electrodes tend to be more correlated than data from distant electrodes. For example, random EEG noise at the Fz electrode will spread to Cz more than to Pz, and the correlation between the data at Fz and the data at Cz will be greater than the correlation between Fz and Pz. In addition, ANOVA is not very robust when the homogeneity of covariance assumption is violated. Violations will lead to artificially low p-values, such that you might get a p-value of less than .05 even when the actual probability of a type I error is 15 percent. Thus, it is important to address violations of this assumption.

The most common way to address this is to use the Greenhouse-Geisser epsilon adjustment (see Jennings & Wood, 1976), which counteracts the inflation of type I errors produced by heterogeneity of variance and covariance (which is also called *nonsphericity*). For each F-value in an ANOVA that has more than 1 degree of freedom in the numerator, this procedure adjusts the degrees of freedom downward—and hence the p-value upward—in a manner that reflects the degree of nonsphericity for each F-value. Fortunately, most major statistics packages provide this adjustment, and it is therefore easy to use.

We used the Greenhouse-Geisser adjustment in the statistical analysis of the P3 amplitude data discussed above. It influenced only the main effects and interactions involving the electrode factors, because the other factors had only two levels (i.e., frequent vs. rare and bright vs. dim). For most of these F-tests, the adjustment didn't matter very much because the unadjusted effects were either not significant to begin with or were so highly significant that a moderate adjustment wasn't a problem (e.g., an unadjusted p-value of .00005 turned into an adjusted p-value of .0003). However, there were a few cases in which a previously significant p-value was no longer significant. For example, in the analysis of the normalized ANOVA, the main effect of anterior-posterior electrode site was significant before the adjustment was applied ($F(2, 18) = 4.37$, p = .0284) but was no longer significant after the adjustment (p = .0586). This may seem like it's not such a great thing, because this effect is no longer significant after the adjustment. However, the original p-value was not accurate, and the adjusted p-value is closer to the actual probability of a type I error. In addition, when very large sets of electrodes are used, the adjustments are usually much larger, and spurious results are quite likely to yield significant p-values when no adjustment is used.

In my opinion, it is absolutely necessary to use the Greenhouse-Geisser adjustment—or something comparable[2]—whenever there are more than two levels of a factor in an ANOVA, especially when one of the factors is electrode site. And this is not just my opinion, but reflects the consensus of the field. Indeed, the journal *Psychophysiology* specifically requires that authors either use an adjustment procedure or demonstrate that their data do not violate the sphericity assumption. In addition, most ERP-savvy reviewers at other journals will require an adjustment procedure. Even if you could "get away" with not using the adjustment, you should use it anyway, because without it your p-values may be highly distorted.

Follow-Up Comparisons

Once you find a significant effect in an ANOVA, it is often necessary to do additional analyses to figure out what the effect means. For example, if you have three experimental conditions and the

ANOVA indicates a significant difference among them, it is usually necessary to conduct additional analyses to determine which of these three conditions differs from the others. Or, if you have a significant interaction, you may need to conduct additional analyses to determine the nature of this interaction. In our P3 experiment, for example, we found an interaction between stimulus brightness and stimulus probability, with a larger effect of probability for the bright stimuli than for the dim stimuli. One could use additional analyses to ask whether the probability effect was significant only for the bright stimuli and not for the dim stimuli, or whether the brightness effect was significant only for the rare stimuli and not for the frequent stimuli.

The simplest way to answer questions such as these is to run additional ANOVAs on subsets of the data. For example, we ran an ANOVA on the data from the bright stimuli and found a significant main effect of probability; we also ran an ANOVA on the data from the dim stimuli and again found a significant main effect of probability. Thus, we can conclude that probability had a significant effect for both bright and dim stimuli even though the original ANOVA indicated that the probability effect was bigger for bright stimuli than for dim stimuli.

When a subset of the data are analyzed in a new ANOVA, it is possible to use the corresponding error term from the original ANOVA (this is called "using the pooled error term"). The advantage to using the pooled error term is that it has more degrees of freedom, leading to greater power. The disadvantage is that any violations of the homogeneity of variance and covariance assumptions will invalidate this approach. Because ERP data almost always violate these assumptions, I would recommend against using the pooled error term in most cases.

There are several additional factors that you must consider when conducting additional analyses of this nature. For example, the more p-values you compute, the greater is the chance that one of them will be significant by chance. However, these additional factors are no different in ERP research than in other areas, and I will not try to duplicate the discussions of these issues that appear in standard statistics texts.

Analyzing Multiple Components

In the standard approach to ERP analysis, a separate ANOVA is conducted for each peak that is measured. It would be possible to include data from separate peaks, with peak as a within-subjects ANOVA factor, but this would not be in the univariate "spirit" of the ANOVA approach. That is, the different components are not really measurements of the same variable under different conditions. Of course, performing a separate ANOVA for each peak can lead to a proliferation of p-values, increasing your experimentwise error, and this can be a substantial problem if you are measuring many different peaks. As chapter 2 discussed, it is usually best to focus an experiment on just one or two components rather than "fishing" for effects in a large set of components, and this minimizes the experimentwise error. It is also possible to enter the data from all of the peaks into a single MANOVA, but multivariate statistics are beyond the scope of this book (for more information, see Donchin & Heffley, 1978; Vasey & Thayer, 1987).

When it is necessary to analyze the data from multiple peaks, it is important to avoid the problem of temporally overlapping components as much as possible. For example, the effects of stimulus probability on the P2, N2, and P3 waves shown in figure 6.5 overlap with each other, especially at the central and parietal electrode sites. In particular, the elimination of the P2 effect at Cz and Pz for the dim stimuli could be due to an increase in P2 latency, pushing the P2 wave into the latency range of the N2 wave and leading to cancellation of the P2 wave by the N2 wave.

When you are faced with data such as these, you really have three choices. The first is to simply ignore the P2 and N2 waves and focus on the P3 wave, which is so much larger than the P2 and N2 waves that it is unlikely to be substantially distorted by them. This is a very reasonable strategy when the experiment was designed to test a specific hypothesis about a single component and the other components are irrelevant. A second approach is to use a more sophisticated approach such as PCA or ICA. This can also be a reasonable approach, as long as you are careful to consider the assumptions that lie behind these techniques. The third approach is to use simple measurement and analysis techniques, but with measurement windows and electrode locations designed to minimize the effects of overlapping components (however, you must be cautious about the possibility that the overlapping components might still distort the results).

As an example of this third approach, we measured P2 amplitude as the mean voltage between 125 and 275 ms, which spanned most of the P2 wave for both the bright and dim stimuli. In addition, to minimize the effects of the overlapping components, we measured the P2 wave only at the frontal electrode sites, where the P2 wave was fairly large and the other components were fairly small. In this analysis, we found a significant main effect of stimulus probability, $F(1,9) = 6.13$, p $< .05$, a significant main effect of stimulus brightness, $F(1,9) = 22.12$, p $< .002$, and a significant probability $\times$ brightness interaction, $F(1,9) = 5.44$, p $< .05$. Although we can't be 100 percent certain that these effects aren't distorted somewhat by the overlapping N2 wave and the effects of stimulus brightness on P2 latency, they correspond well with the waveforms shown in figure 6.5 and are probably real.

The Bottom Line

In this section, I have discussed the most common approach to performing statistical analyses on ERP data. As I mentioned at the beginning of the section, this approach is not without flaws, but it also has two important positive attributes. First and foremost, it is a conventional approach, which means that people can easily understand and evaluate results that are analyzed in this manner. Second, even though a rigid .05 criterion is somewhat silly when the assumptions of ANOVA are violated and when a large number of p-values are calculated, it is reasonably conservative and allows everyone to apply a common standard. My ultimate recommendation, then, is to use the standard approach when presenting results to other investigators, but to rely on high levels of statistical power, replication, and common sense when deciding for yourself whether your results are real.

Suggestions for Further Reading

The following is a list of journal articles and book chapters that provide useful information about ERP measurement and analysis techniques.

Coles, M. G. H., Gratton, G., Kramer, A. F., & Miller, G. A. (1986). Principles of signal acquisition and analysis. In M. G. H. Coles, E. Donchin & S. W. Porges (Eds.), *Psychophysiology: Systems, Processes, and Applications* (pp. 183–221). New York: Guilford Press.

Donchin, E., & Heffley, E. F., III. (1978). Multivariate analysis of event-related potential data: A tutorial review. In D. Otto (Ed.), *Multidisciplinary Perspectives in Event-Related Brain Potential Research* (pp. 555–572). Washington, D.C.: U.S. Government Printing Office.

Jennings, J. R., & Wood, C. C. (1976). The e-adjustment procedure for repeated-measures analyses of variance. *Psychophysiology*, *13*, 277–278.

Picton, T. W., Bentin, S., Berg, P., Donchin, E., Hillyard, S. A., Johnson, R., Jr., Miller, G. A., Ritter, W., Ruchkin, D. S., Rugg, M. P., & Taylor, M. J. (2000). Guidelines for using human event-related potentials to study cognition: recording standards and publication criteria. *Psychophysiology*, *37*, 127–152.

Rosler, F., & Manzey, D. (1981). Principal components and varimax-rotated components in event-related potential research: Some remarks on their interpretation. *Biological Psychology*, *13*, 3–26.

Ruchkin, D. S., & Wood, C. C. (1988). The measurement of event-related potentials. In T. W. Picton (Ed.), *Human Event Related Potentials* (pp. 121–137). Amsterdam: Elsevier.

Squires, K. C., & Donchin, E. (1976). Beyond averaging: The use of discriminant functions to recognize event related potentials elicited by single auditory stimuli. *Electroencephalography and Clinical Neurophysiology, 41,* 449–459.

Urbach, T. P., & Kutas, M. (2002). The intractability of scaling scalp distributions to infer neuroelectric sources. *Psychophysiology, 39,* 791–808.

Vasey, M. W., & Thayer, J. F. (1987). The continuing problem of false positives in repeated measures ANOVA in psychophysiology: A multivariate solution. *Psychophysiology, 24,* 479–486.

Wood, C. C., & McCarthy, G. (1984). Principal component analysis of event-related potentials: Simulation studies demonstrate misallocation of variance across components. *Electroencephalography and Clinical Neurophysiology, 59,* 249–260.

The ultimate goal of cognitive neuroscience is to understand how the neural circuitry of the brain gives rise to cognitive processes. This is a challenging enterprise, and one of the central difficulties is to measure how specific populations of neurons operate during the performance of various cognitive tasks. The best techniques for measuring activity in specific populations of neurons are invasive and cannot usually be applied to human subjects, but it is difficult to study many aspects of cognition in nonhuman subjects. PET and fMRI provide noninvasive means of localizing changes in blood flow that are triggered by overall changes in neural activity, but blood flow changes too slowly to permit the measurement of most cognitive processes in real time. ERPs provide the requisite temporal resolution, but they lack the relatively high spatial resolution of PET and fMRI. However, ERPs do provide some spatial information, and many investigators are now trying to use this spatial information to provide a measurement of the time course of neural activity in specific brain regions.

The goal of this chapter is to explain how this process of ERP *source localization* works in general and to provide a discussion and critique of the most common source localization techniques. Before I begin, however, I would like to provide an important caveat: I tend to be extremely skeptical about ERP localization, and this chapter reflects that skepticism. Many other researchers are also skeptical about ERP localization (see, e.g., Snyder, 1991), but I am more skeptical than most. Consequently, you should not assume that my conclusions and advice in this chapter represent those of the majority of ERP researchers. You should talk to a broad range of ERP experts before making up your mind about source localization.

Source localization is very complex, both in terms of the underlying mathematics and the implementation of the procedures, and it would require an entire book to provide a detailed description of the major techniques. This chapter therefore focuses on providing a simple description of the two major classes of techniques—including their strengths and weaknesses—so that you can understand published research using these techniques and decide whether to pursue localization yourself.

My main advice in this chapter is that beginning and intermediate ERP researchers should not attempt to localize the sources of their ERP data. ERP localization is a very tricky business, and doing it reasonably well requires sophisticated techniques and lots of experience. Moreover, the techniques currently in wide use are not completely satisfactory. However, many smart people are working to improve the techniques, and some promising approaches are on the horizon. Thus, you may eventually want to do some source localization, and you will certainly be reading papers that report localization results. The goal of this chapter is therefore to make you an informed and critical observer of source localization efforts rather than a participant in these efforts.

The Big Picture

If you place a single dipole in a conductive sphere, you can use relatively simple equations to predict the precise distribution of observable voltage on the surface of the sphere. This is called the *forward problem*, and it is relatively easy to solve. Voltages summate linearly, which means that the forward problem is also easy to solve for multiple simultaneously active dipoles—the voltage distributions for the individual dipoles are simply added together to derive the distribution for the set of dipoles. The forward problem can also be solved for realistic head shapes.

The problem arises in solving the *inverse problem* of determining the positions and orientations of the dipoles on the basis of the observed distribution of voltage over the scalp. If only one dipole is present, and there is no noise, then it is possible to solve the inverse problem to any desired degree of spatial resolution by comparing forward solutions from a model dipole with the observed scalp distribution and then adjusting the dipole to reduce the discrepancy between the predicted and observed distributions. However, it is not possible to solve the inverse problem if you don't know the number of dipoles (or if the activity is distributed rather than dipolar) because there is no unique solution to the inverse problem in this case. In other words, for any given scalp distribution, there is an infinite number of possible sets of dipoles that could produce that scalp distribution (Helmholtz, 1853; Plonsey, 1963). Thus, even with perfectly noise-free data, there is no perfect solution to the inverse problem.

Several investigators have proposed ways around this uniqueness problem, and their solutions fall into two general categories. One approach is to use a small number of equivalent current dipoles, each of which represents the summed activity over a small cortical region (perhaps 1–2 cm^3), and assume that these dipoles vary only in strength over time; this is the *equivalent current dipole* category of source localization methods. The second category divides the brain's volume (or the cortical surface) into a fairly large number of voxels (perhaps a few thousand), and computes the set of strengths for these voxels that can both explain the observed distribution of voltage over the scalp and satisfy additional mathematical constraints; this is the *distributed source* category.

The Forward Solution

Early forward solutions assumed that the head was a sphere, which makes the computations relatively simple. That is, if we assume that the head is a sphere and that the head's resistance is homogeneous, it is very easy to compute the voltage created at every point on the scalp by a single dipole or an arbitrarily large set of dipoles. The skull and scalp have a higher resistance than the brain, but this is easily accommodated by a model in which skull and scalp layers cover a homogeneous, spherical brain. Researchers often use a spherical approximation for computing forward solutions because it is easy to generate the model and because they can compute the forward solution very rapidly. However, this model is obviously an oversimplification, and many researchers are now using more detailed models of the head, such as *finite element models*. A finite element model divides the volume of the head into thousands or even hundreds of thousands of individual cubes. The resistance within each cube is assumed to be homogeneous (which is a reasonable approximation given that each cube is very small), but the resistances are different for each cube. In addition, the set of cubes that define the head are not constrained to form a large sphere, but instead take into account the shape of each subject's head (as determined by a structural MRI scan).

Once the researcher has divided the head into a set of cubes and determined the resistances of the individual cubes, simple equations can determine the surface voltages that a dipole in a known location would produce. Given the large number of cubes in a realistic model of the head, it takes a lot of computer power to do this sort of modeling, but as computers become faster, this sort of modeling will fall within the means of more and more researchers. The hard part is to determine the resistances of the individual cubes. The resistances can't actually be measured for a living human subject, but they can be estimated by performing an MRI scan, dividing up the head into regions of different tissue types on the basis of the MRI scan, and using normative resistance values for each tissue type (i.e., resistance values that have been determined invasively from cadavers, from nonhuman animals, and possibly from human neurosurgical patients).

Because finite element models are so computationally intensive, researchers have developed a less intensive variant, the *boundary element model*. This approach takes advantage of the fact that the brain itself has a relatively constant resistance, and most of the action is in the boundaries between the brain, the skull, and the scalp. The model therefore consists of boundary surfaces for these tissues, with the assumption that the resistance within a tissue is constant throughout the extent of that tissue. As in finite element models, the boundaries are estimated from structural MRI scans and the conductivities of each tissue type are based on normative values.

Equivalent Current Dipoles and the BESA Approach

The *brain electrical source analysis* (BESA) technique is the prototypical example of an equivalent current dipole technique. In addition, BESA has been the most commonly used method for localizing ERP sources over the past twenty years, in part because it is relatively simple and inexpensive (both in terms of computing resources and money). Thus, this section will focus on the BESA technique. The interested reader may also want to learn about the MUSIC (multiple signal characterization) technique, which provides a more sophisticated approach to computing equivalent current dipole solutions (see Mosher, Baillet, & Leahy, 1999; Mosher & Leahy, 1999; Mosher, Lewis, & Leahy, 1992).

BESA is based on the assumption that the spatiotemporal distribution of voltage can be adequately modeled by a relatively small set of dipoles (< 10), each of which has a fixed location and orientation but varies in magnitude over time (Scherg, Vajsar, & Picton, 1989; Scherg & von Cramon, 1985). Each dipole has five major parameters, three indicating its location, and two indicating its orientation. A magnitude parameter is also necessary, but this parameter varies over time and is treated differently from the location and orientation parameters.

The Essence of BESA

The BESA algorithm begins by placing a set of dipoles in an initial set of locations and orientations, with only the magnitude being unspecified. The algorithm then calculates a forward solution scalp distribution for these dipoles, computing a magnitude for each dipole at each point in time such that the sum of the dipoles yields a scalp distribution that fits, as closely as possible, the observed distribution for each point in time.

The scalp distributions from the model are then compared with the observed scalp distributions at each time point to see how well they match. The degree of match is quantified as the percentage of the variance in scalp distribution that is explained by the model; alternatively, it can be expressed as the percentage of unexplained variance (called the *residual variance*). The goal of the algorithm is to find the set of dipole locations and orientations that yields the lowest residual variance, providing the best fit between the model and the data.

This is accomplished in an iterative manner, as shown in figure 7.1. On each iteration, the forward solution is calculated, leading to a particular degree of residual variance, and then the positions and orientations of the dipoles are adjusted slightly to try to reduce the residual variance. This procedure is iterated many times using a gradient descent algorithm so that the positions and orientations will be adjusted in a way that tends to decrease the residual variance with each successive iteration. In the first several iterations, the residual variance drops rapidly, but after a large number of iterations, the residual variance stops declining much from one iteration to the next and the dipole positions and orientations become stable. There are various refinements that one can add, but this is the essence of the BESA technique.

Figure 7.2 shows an example of a BESA solution (from Di Russo et al., 2002). The goal of this source localization model was to characterize the generators of the early visual sensory components. The top of the figure shows the scalp distribution of the ERP response to a checkerboard stimulus in the upper left visual field in various

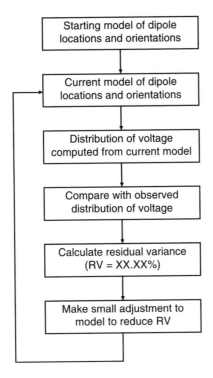

Figure 7.1 Basic procedure of the BESA technique.

time ranges, and the bottom of the figure shows the BESA model that was obtained. Each dipole is represented by a dot showing its location and a line showing its orientation; three different views of the head are shown so that the dipoles can be seen in three dimensions. Each dipole is also associated with a *source wave-form*, which shows how the estimated magnitude for that dipole varies over time.

The Starting Point

Before initiating this iterative procedure, it is necessary to decide how many dipoles to use and what their starting positions will be. These are important decisions, and they will have a very

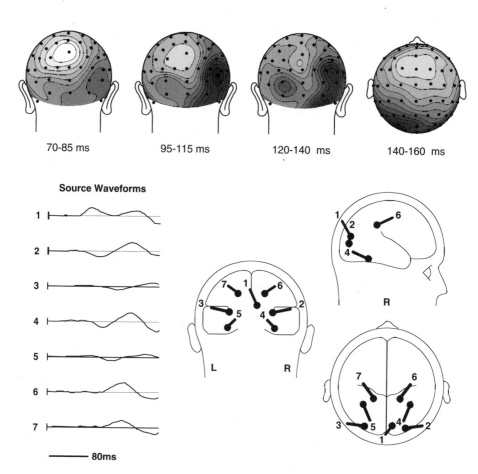

Figure 7.2   Example of an equivalent current dipole model generated using the BESA technique (from the study of Di Russo et al., 2002). The top shows the scalp distributions of voltage measured in various time ranges in response to a checkerboard in the upper left visual field. This time-space-voltage data set is modeled by a set of seven dipoles. The locations and orientations of the dipoles are shown in three different views in the lower right region, and the magnitude of each dipole over time is shown in the lower left region. (© 2002 Wiley-Liss, Inc.) Thanks to Francesco Di Russo for providing electronic versions of these images.

substantial impact on the solution the algorithm ultimately reaches. In fact, the most problematic aspect of the BESA technique is the fact that the user has a lot of control over the results (this is called *operator dependence*); consequently the results may be biased by the user's expectations.

You can use several different strategies to select the number of dipoles. One approach is to use principal components analysis (PCA) to determine how many underlying spatial distributions of activity contribute to the observed distribution of voltage over the scalp. This technique can work well in some cases. However, if the magnitudes of two dipoles are correlated with each other over time, they may be lumped together into a single component. Thus, the number of components identified by PCA can only provide a lower bound on the number of dipoles.

Another strategy is to start by using one or two dipoles to create a model of the early part of the waveform, under the assumption that the response begins in one or two sensory areas. Once a stable solution is reached for the early part of the waveform, the time window is increased and new dipoles are added while the original dipoles are held constant. This procedure is then repeated with a larger and larger time window. This procedure makes some sense, but it has some shortcomings. First, at least a dozen visual areas are activated within 60 ms of the onset of activity in visual cortex (Schmolesky et al., 1998), so representing the early portion of the waveform with one or two dipoles is clearly an oversimplification. Second, some error is likely in any dipole solution, and small errors in the initial dipoles will lead to larger errors in the next set of dipoles, and the location estimates will become increasingly inaccurate as you add more and more dipoles.

A third strategy is to use preexisting knowledge about the brain to determine the number of dipoles. For example, if you use difference waves to isolate the lateralized readiness potential, it would be reasonable to start with the assumption that two dipoles are present, one in each hemisphere.

There are also several strategies that you can use for determining the starting positions and orientations of the dipoles, and different strategies will lead to different results. Unfortunately, most papers using the BESA technique describe the model produced by one or two different starting positions. Almost every time I have read a paper that used the BESA technique, I wished the authors had provided a detailed description of the results that would have been obtained with a wide variety of different starting locations. It would be possible for researchers to do this. Indeed, Aine, Huang, and their colleagues have developed what they call a *multi-start* approach to localization, in which they apply the localization algorithm hundreds or even thousands of times with different starting parameters (Aine et al., 2000; Huang et al., 1998). It is then possible to determine which dipole locations occur frequently in the solutions and are therefore relatively independent of the starting parameters.

Another strategy is to start with dipoles in locations that are based on preexisting knowledge about the brain. The locations could be based on general knowledge (e.g., the location of primary and secondary auditory areas), or they could be based on specific results from previous experiments (e.g., activation centers from a similar fMRI experiment). When using the latter approach, researchers sometimes say that the solution is based on *seeded dipoles*, and some studies have explicitly shown that similar results were obtained with random dipole locations and seeded dipoles (e.g., Heinze et al., 1994). This is, in some ways, a reasonable approach. However, it seems likely to lead to a confirmation bias, increasing the likelihood that the expected results will be obtained even if they are not correct.

Shortcomings of the BESA Approach

The BESA approach has several shortcomings, but the most significant problem is that there is no mathematically principled means of quantifying the accuracy of a solution. Specifically, in the presence of noise, it is possible for a substantially incorrect solution to have the same (or lower) residual variance than the correct solution. Even with minimal noise, it is possible for a substantially incorrect solution to have a very low residual variance (especially when using more than a few dipoles). One reason for this is that each BESA dipole has five free parameters (plus a time-varying magnitude parameter). Thus, a model with only six dipoles has thirty free parameters, and a relatively large error in one of these parameters can easily be offset by small adjustments in the other parameters, resulting in low residual variance. Even if only one dipole is present, the BESA solution may be inaccurate due to noise in the data and errors in the head model. Without some means of quantifying the likelihood that a solution is correct or even nearly correct, it's hard to use a BESA solution to provide strong support for or against a hypothesis.

The second most significant shortcoming of the BESA technique is the operator dependence of the technique (as mentioned briefly in the previous section). In addition to setting the number and initial positions of the dipoles, a researcher can adjust several other parameters that control how the algorithm adjusts the positions and orientations of the dipole while searching for the configuration with the least residual variance. Moreover, at several points in the process, the researcher makes subjective decisions about adding or deleting dipoles from the solution or changing various constraints on the dipoles. I have seen ERP researchers spend weeks applying the BESA technique to a set of data, playing around with different parameter settings until the solution "looks right." Of course, what "looks right" is often a solution that will confirm the researcher's hypothesis (or at least avoid disconfirming it).

Another significant shortcoming of the BESA technique is that it will produce an incorrect solution if the number of dipoles is incorrect. It is difficult or impossible to know the number of dipoles in advance, especially in an experiment of some cognitive complexity, so this is a significant limitation. Moreover, BESA uses a discrete dipole to represent activity that may be distributed across a fairly large region of cortex, and this simplification may lead to substantial errors.

A Simulation Study

There have been a variety of tests of the accuracy of equivalent current dipole localizations, but they have mostly used only one or two simultaneously active dipoles (see, e.g., Cohen & Cuffin, 1991; Leahy et al., 1998). These simulations are useful for assessing the errors that might be likely in very simple sensory experiments, but they do not provide meaningful information about the errors that might occur in most cognitive neuroscience experiments.

Miltner et al. (1994) performed the most informative simulation study in the context of cognitive neuroscience. This study used BESA's spherical, three-shell head model to simulate a set of dipoles and produce corresponding ERP waveforms from thirty-two electrode sites. From these ERP waveforms, nine participants attempted to localize the dipoles using BESA. The participants consisted of ERP researchers with various levels of expertise with BESA (including three with very high levels of expertise).[1] The participants were told that the data were simulated responses from left somatosensory stimuli that were presented as the targets in an oddball task, and they were given the task of trying to localize the sources. The simulation comprised ten dipoles, each of which was active over some portion of a 900-ms interval. White noise was added to the data to simulate the various sources of noise in real ERP experiments. The simulation included two dipoles in left primary somatosensory cortex (corresponding to the P100 wave and an early portion of the N150 wave), a mirror-symmetrical pair of dipoles in left and right secondary somatosensory cortex, midline dipoles in prefrontal and central regions, and mirror-symmetrical pairs of dipoles in medial temporal and dorsolateral prefrontal regions.

This is a fairly large set of dipoles,[2] but the participants' task was made easier by at least seven factors: (1) the solution included several dipoles that were located exactly where they would be expected (e.g., the primary and secondary somatosensory areas); (2) three of the dipole pairs were exactly mirror-symmetrical (which matches a typical BESA strategy of assuming mirror symmetry at the early stages of the localization process); (3) the spherical BESA head model was used to create the simulations, eliminating errors due to an incorrect forward solution; (4) the temporal overlap between the different dipoles was modest (for most of the dipoles, there was a time range in which only it and one other dipole or mirror-symmetrical dipole pair were strongly active); (5) with the exception of the two dipoles in primary somatosensory cortex, the dipoles were located fairly far away from each other; (6) the white noise that was added was more easily filtered out than typical EEG noise and was apparently uncorrelated across sites; and (7) the simulation used discrete dipoles rather than distributed regions of activation.

Despite the fact that the simulation perfectly matched the assumptions of the BESA technique and was highly simplified, none of the participants reached a solution that included all ten dipoles in approximately correct positions. The number of dipoles in the solutions ranged from six to twelve, which means that there were several cases of missing dipoles and/or spurious dipoles. Only two of the nine participants were able to distinguish between the midline prefrontal and midline central dipoles, and the other seven participants tended to merge them into a single dipole even though the actual dipoles were approximately 5 cm apart.

Across all dipoles that appeared to be localized by the participants, the average localization error was approximately 1.4 cm, which doesn't sound that bad. However, this was a simplified simulation based on the BESA head model, and the errors with real data are likely to be greater. Moreover, there were many cases in which an individual dipole's estimated location was 2–5 cm from the actual dipole's location, and the mean errors across dipoles for individual participants were as high as 2 cm. To be fair, however, I should note that most of the participants provided a reasonably accurate localization of one of the two primary somatosensory dipoles, the secondary somatosensory dipoles, the medial temporal lobe dipoles, and the dorsolateral prefrontal dipoles. But each of the nine participants had at least one missing dipole, one spurious dipole, or one mislocalization of more that 2 cm.

From this study, we can draw two main conclusions. First, in this highly simplified situation, dipoles were often localized with a reasonable degree of accuracy, with an average error of 1–2 cm for most of the participants (relative to a 17-cm head diameter). Thus, when reality does not deviate too far from this simplified situation, the BESA technique can potentially provide a reasonable estimate of the locations of most of the dipoles most of the time. However, some of the simplifications seem quite far from reality, so it is entirely possible that average errors will be considerably larger with most real data sets.

The second main conclusion is that any single dipole in a given multiple-dipole BESA model has a significant chance of being substantially incorrect, even under optimal conditions. Dipoles may be mislocalized by several centimeters or completely missed; multiple dipoles may be merged together, even if they are fairly far apart; and spurious dipoles may be present in the model that correspond to no real brain activity. Thus, even if the average error is only 1–2 cm for most dipoles, this simulation suggests that BESA solutions for moderately complex data sets may typically contain at least one missing dipole, one spurious dipole, or one 2–5 cm localization error. And the accuracy of the technique is presumably even worse for real data sets that deviate from the simplifications of this simulation.

Although the BESA technique has been widely used over the past twenty years, most ERP researchers now appreciate its limitations. There is a clear trend away from this technique and toward more sophisticated equivalent current dipole approaches and distributed source approaches.

Distributed Source Approaches

General Approach

Instead of using a small number of equivalent current dipoles to represent the pattern of neural activity, it is possible to divide the brain up into a small number of voxels and find a pattern of activation values that will produce the observed pattern of voltage on the surface of the scalp. For example, you could divide the surface of the brain into a hundred little cubes. Each cube would contain three dipoles, one pointing upward, one pointing forward, and one pointing laterally (a single dipole of an arbitrary orientation can be simulated by varying the relative strengths of these three dipoles). You could then find a pattern of dipole strengths that would yield the observed distribution of voltage on the surface of the head. This would provide you with an estimate of the distribution of electrical activity throughout the brain.

The problem with this approach is that even this relatively coarse parcellation of the brain requires that you compute 300 different dipole strengths. That is, your model has 300 free parameters to be estimated. Generally speaking, you need at least as many independent data points as you have free parameters, and even if you have voltage measurements from 300 electrodes, they are contaminated by noise and are not independent of each other. Consequently, there are many different sets of strengths of the 300 dipoles that could produce the observed ERP scalp distribution. This is the problem of *nonuniqueness*. And it would get even worse if we wanted to divide the brain into even smaller voxels.

Cortically Constrained Models

Researchers have developed several strategies to avoid the nonuniqueness problem. One strategy is to reduce the number of dipoles by assuming that scalp ERPs are generated entirely by currents generated in the cerebral cortex, flowing perpendicular to the cortical surface (which is a reasonable assumption in most cases). Instead of using a set of voxels that fills the entire volume of the brain, with three dipoles per voxel, this approach uses structural MRI scans to divide the cortical surface into hundreds or thousands of small triangles, each with a single dipole oriented perpendicular to the cortical surface. This *cortically constrained* approach dramatically reduces the number of free parameters in the model (although some error may be introduced by inaccuracies in the cortical surface reconstruction). The result is a model of the distribution of electrical activity over the cortical surface.

Figure 7.3 illustrates this approach, with a slice through a cartoon brain that shows the cortical surface and recording electrodes for the left hemisphere. The cortical surface has been divided into a number of small patches, and each patch is treated as a dipolar current source pointing perpendicular to the cortical surface. This reduces the number of dipole locations and orientations compared to dividing the entire volume of the brain into voxels, each of which contains three dipoles. However, the number of dipoles needed in a real experiment is still very large (usually in the hundreds or thousands), and there is still no unique pattern of dipole strengths that can account for the observed distribution of voltage on the scalp. That is, the use of a cortically constrained model reduces the number of internal patterns of activity that could explain the observed distribution of voltage over the scalp, but it does not bring the number all the way to one (i.e., to a unique solution).

The non-uniqueness problem in cortically constrained models can be appreciated by considering sources 15 and 16 in figure 7.3. These sources are almost perfectly parallel to each other, but they are inverted in orientation with respect to each other (i.e., the outer surface of the cortex points downward for source 15 and upward for source 16). This is a common occurrence given the extensive foldings of the human cerebral cortex. The non-uniqueness problem occurs because any increase in the magnitude of source 15

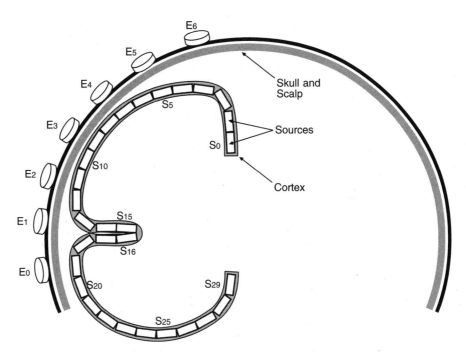

Figure 7.3 Example of the electrical sources and measurement electrodes used by cortically constrained distributed source localization methods. The figure shows a coronal section through the brain. The cortex is divided into a large number of patches that are assumed to be the electrical sources (labeled S_0–S_{29} here). Each source is modeled as a dipole that is centered in the corresponding cortical patch and oriented perpendicular to the patch. The voltage corresponding to each source propagates through the brain, skull, and scalp to reach the recording electrodes (labeled E_0–E_6), and the voltages from the different sources simply sum together.

can be cancelled by an increase in the magnitude of source 16, with no change in the distribution of voltage on the surface.

The Minimum Norm Solution

To get around the nonuniqueness problem (for both whole-brain and cortically constrained models), Hämäläinen and Ilmoniemi (1984) proposed adding an additional constraint to the system. This constraint is based on the fact that the cancellation problem—as exemplified by sources 15 and 16 in figure 7.3—allows the magnitudes of nearby sources to become huge without distorting the distribution of voltage on the scalp. These huge magnitudes are a biologically unrealistic consequence of the modeling procedure, and it therefore makes sense to eliminate solutions that have huge magnitudes. Thus, Hämäläinen and Ilmoniemi (1984) proposed selecting the one solution that both produces the observed scalp distribution and has the minimum overall source magnitudes. This is called the *minimum norm* solution to the problem of finding a unique distribution of source magnitudes.

One shortcoming of the minimum norm solution is that it is biased toward sources that are near the surface, because a larger magnitude is necessary for a deep source to contribute as much voltage at the scalp as a superficial source. However, this problem can be solved by using a *depth-weighted minimum norm* solution that weights the magnitudes of each source according to its depth when finding the solution with the minimum overall source magnitudes.

Other researchers have proposed other types of minimum norm solutions that reflect different constraints. The most widely used of these alternatives is the low-resolution electromagnetic tomography (LORETA) technique, which assumes that the voltage will change gradually (across the volume of the brain or across the cortical surface) and selects the distribution of source magnitudes that is maximally smooth (Pascual-Marqui, 2002; Pascual-Marqui et al., 2002; Pascual-Marqui, Michel, & Lehmann, 1994). The smoothness constraint may be reasonable in many cases, but sharp borders exist between adjacent neuroanatomical areas, and these borders would sometimes be expected to lead to sudden changes in cortical current flow. Indeed, if an experimental manipulation is designed to activate one area (e.g., V3) and not an adjacent area (e.g., V4), then the goals of the experiment would be incompatible with an assumption of smoothness. On the other hand, gradual changes in activity are probably the norm within a neuroanatomical area, so the smoothness constraint may be appropriate in many cases. Note also that, because of its smoothness constraint, the LORETA technique is appropriate only for finding the center of an area of activation and not for assessing the extent of the activated area. In contrast, nothing about the original and depth-weighted minimum norm solutions will prevent sharp borders from being imaged.

It is possible to combine empirical constraints with these mathematical constraints. For example, Dale and Sereno (1993) describe a framework for using data from functional neuroimaging to provide an additional source of constraints that can be combined with the minimum norm solution (see also George et al., 1995; Phillips, Rugg, & Friston, 2002; Schmidt, George, & Wood, 1999). Like the LORETA approach, this approach has the advantage of using biological information to constrain which solution is chosen. However, it is easy to conceive of situations in which an fMRI effect would not be accompanied by an ERP effect or vice versa (see Luck, 1999), so the addition of this sort of neuroimaging-based constraint may lead to a worse solution rather than a better one. It may also produce a confirmation bias: when you use fMRI data to constrain your ERP localization solution, you're increasing the likelihood of finding a match between the ERP data and the fMRI data even if the ERP is not generated at the locus of the fMRI BOLD signal. A simulation study suggested that the most problematic situation arises when an ERP source is present without a corresponding fMRI source (Liu, Belliveau, & Dale, 1998). This study also indicated that the resulting distortions are reasonably small if the source localization algorithm assumes a less-than-perfect correspondence between the ERP and fMRI data. It remains to be seen whether the use of fMRI data to probabilistically constrain ERP source localization leads to substantial errors when applied to real data.

Unlike equivalent source dipole approaches, minimum norm-based techniques will always find a unique solution to the inverse problem, and they do it largely automatically. However, a unique and automatic solution is not necessarily the correct solution. The correctness of the solution will depend on the correctness of the assumptions. As Ilmoniemi (1995) discussed, an approach of this type will provide an optimal solution if its assumptions are valid, but different sets of assumptions may be valid for different data sets. For example, if one applies the LORETA technique to the three-dimensional volume of the brain without first creating a model of the cortical surface, the smoothness assumption will almost certainly be violated when areas that are distant from each other along the cortical surface abut each other due to the folding pattern of the cortex. But if the smoothness constraint is applied along the reconstructed 2-D cortical surface, then this assumes that subcortical regions do not contribute to the data, which may be incorrect.

The Added Value of Magnetic Recordings

As described in chapter 1, the EEG is accompanied by a magnetic signal, the MEG, and event-related electrical potentials (ERPs) are accompanied by event-related magnetic fields (ERMFs). Because the skull is transparent to magnetism, it does not blur the MEG signal, and this leads to improved spatial resolution for MEG recordings. Another benefit of MEG is that, because magnetism passes unimpeded through the head, MEG/ERMF localization does not require a model of the conductances of the head; it simply requires a model of the overall shape of the brain. Thus, it can be advantageous to apply localization techniques to ERMFs rather than ERPs.

ERMF localization faces the same non-uniqueness problem as ERP localization, but combining ERP and ERMF data provides a new set of constraints that can aid the localization process. The main reason for this is that the voltage field and the magnetic field run in different directions, and they therefore provide complementary information. As figure 7.4A illustrates, the magnetic field runs in circles around the current dipole. When the current dipole is oriented in parallel to the skull, the magnetic field exits the skull on one side of the dipole and reenters the skull on the other side (figure 7.4B). The strength of the magnetic field varies as a function

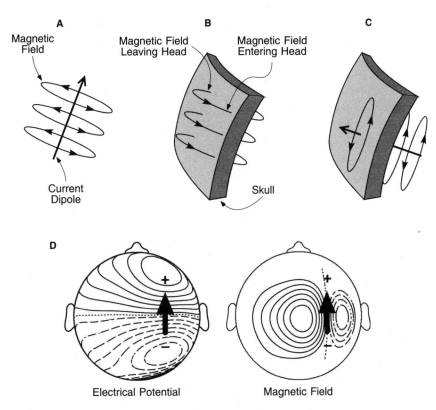

Figure 7.4 Relationship between an electrical dipole and its associated magnetic field. An electrical dipole has a magnetic field running around it (A), and when the dipole is roughly parallel to the surface of the head, the magnetic field leaves and reenters the head (B). If the dipole is oriented radially with respect to the head, the magnetic field does not vary across the surface of the head (C). When a dipole runs parallel to the surface of the head (represented by the arrow in D), there is a broad region of positive voltage at the positive end (solid lines) and a broad region of negative voltage at the negative end (dashed lines), separated by a line of zero voltage (represented by the dotted line). The magnetic field, in contrast, consists of magnetic flux leaving the head on one side of the dipole (solid lines) and reentering the head on the other side (dashed lines), separated by a line of no net flux (dotted line). Thanks to Max Hopf for providing the electrical and magnetic distributions shown in (D).

of distance from the dipole, just like the strength of the electrical field, but the voltage distribution is broader due to the blurring of the scalp (figure 7.4D). In addition, the magnetic and electrical distributions are oriented at 90 degrees with respect to each other. As figure 7.4D shows, the positive and negative electrical potentials appear at the positive and negative ends of the dipole, and the line of zero voltage runs perpendicularly through the center of the dipole. The efflux and influx of the magnetic field, in contrast, occur on the left and right sides of the dipole, and the zero flux line runs in parallel with the orientation of the dipole.

MEG fields differ from EEG fields in another key way as well. If the current dipole is perfectly perpendicular to the skull, as in figure 7.4C, the magnetic field does not exit and reenter the head, and it is essentially invisible. As the dipole tilts from perpendicular toward parallel, a recordable magnetic field begins to appear again. In contrast, a large and focused voltage will be present directly over a perpendicular dipole. A dipole near the center of the head will act much like a perpendicular dipole, generating a reasonably large voltage on the surface of the scalp that is accompanied by a magnetic field that does not exit and reenter the head and is therefore effectively invisible. Thus, magnetic signals are largest for superficial dipoles that run parallel to the surface of the skull, and fall off rapidly as the dipoles become deeper and/or perpendicularly oriented, but voltages do not fall off rapidly in this manner.

The different effects of dipole depth and orientation on electrical and magnetic signals provide an additional set of constraints on source localization solutions. In essence, there are many internal source configurations that can explain a given electrical distribution, and there are also many internal source configurations that can explain a given magnetic distribution. But there will be far fewer configurations than can explain both the electrical distribution and the magnetic distribution. Consequently, the combination of magnetic and electrical data is substantially superior to either type of data alone. The main drawbacks of combining magnetic and electrical data compared to using magnetic data alone are that (a) a more complex head model is needed for the electrical data, and (b) some effort is required to ensure that the electrical and magnetic data are in exactly the same spatial reference frame.

Can We Really Localize ERPs?

Each of the source localization techniques described in this chapter has shortcomings. Of course, any scientific technique has limitations and shortcomings, but the shortcomings of source localization techniques are fundamentally different from the shortcomings of other techniques for localization of function. This section will explore these differences and consider a new approach that seems more promising.

Source Localization as Model Fitting

To understand the essence of ERP source localization, it is useful to compare it with a "true" neuroimaging technique, such as PET. In the most common PET approach, radioactively labeled water molecules travel through the bloodstream, where their diffusion properties are straightforward. Consequently, the number of radioactive molecules in a given volume of the brain can be directly related to the flow of blood through that part of the brain. When a labeled water molecule decays, it gives off a positron, which travels a known distance (or distribution of distances) before colliding with an electron. This collision leads to a pair of annihilation photons that travel in opposite directions along the same line. When these high-intensity photons are picked up simultaneously by two detectors within the ring of detectors around the subject, there is a high likelihood that they were generated somewhere along the line between the two detectors, and the decaying isotope is known to have been within a certain distance from this line. Thus, by combining the known physics of radiation with various probability distributions, one can directly compute the maximum likelihood location of the radioactively labeled water molecules and the margin of error of this location. The story is analogous, although more complicated, for fMRI.

Because the ERP localization problem is underdetermined, mainstream ERP localization techniques employ a different approach. That is, they do not simply compute the maximum likelihood location of an ERP source, along with a margin of error, on the basis of the physics of electricity and magnetism. Instead, ERP localization techniques generate *models* of the underlying distribution of electrical activity, and these models are evaluated in terms of their ability to satisfy various constraints. The most fundamental constraint, of course, is that a given model must recreate the observed distribution of voltage over the surface of the head. However, a correct model may not fit the data exactly, because noise in the data distorts the observed distribution. Consequently, any internal configuration that is, say, 95 percent consistent with the observed scalp distribution might be considered acceptable. Unfortunately, there will be infinitely many internal configurations that can explain an observed scalp distribution, especially when only a 95 percent fit is required.

Additional constraints are then added to select one internal configuration from the many that can explain the observed scalp distribution. Each source localization technique embodies a different set of these additional constraints, and the constraints can be either mathematical (as in the use of the minimum norm) or empirical (as in the use of fMRI data to constrain ERP localizations). The most straightforward empirical constraint is the use of structural MRI scans to constrain the source locations to the cortical surface. However, this alone does not lead to a unique solution (and it may not always be the case that all scalp ERP activity arises from the cortex). Researchers therefore add other constraints, but there is usually no way of assessing whether these constraints are correct and sufficient.

The bottom line is that ERP localization leads to a *model* of the internal configuration of electrical activity, not a *measurement* of the internal distribution of electrical activity. In contrast, PET and fMRI provide measurements and not merely models. PET, for example, provides a measurement of the internal distribution of radioactively labeled blood. This measurement is derived from more basic measurements, but that is true of most sophisticated scientific measurements. And although the PET measurements are not error-free, this is true of any measurement, and one can specify the margin of error. It is more difficult to describe exactly what the BOLD signal reflects in fMRI, but the location of this signal is measured with a known margin of error. In contrast, one cannot use surface electrodes to measure the distribution of internal electrical activity with a known margin of error.

People occasionally ask me how accurately ERPs can be localized, hoping for a quantification of accuracy that they can compare with the accuracy of PET and fMRI. My response is that the accuracy of ERP localization is simply undefined. That is, in the absence of any constraints beyond the observed scalp distribution, radically different distributions of internal electrical activity would produce the observed scalp distribution, and the margin of error is essentially the diameter of the head.

Once constraints are added, some source localization approaches could, in principle, quantify the margin of error. For example, it would be possible to state that the estimated center of a region of activation is within X millimeters of the actual center of the activated region. Or it would be possible to state that the amount of estimated current flow within each patch of cortex is within Y percent of the actual current flow. I've never seen anyone do this in the context of a serious experiment, but it would be an extremely useful addition to the source localization techniques. However, the margin of error that could be specified in this manner would be meaningful only if the constraints of the model were fully adequate and the only sources of error arose from noise in the ERP data (and perhaps errors in specifying the head model). If the constraints were insufficient, or if they reflected incorrect assumptions about the underlying neural activity, then the margin of error would be meaningless. Thus, the source localization techniques that are currently in widespread use do not, in practice, provide a meaningful estimate of the margin of error.

Probabilistic Approaches

The commonly used source localization techniques attempt to find a single pattern of internal electrical activity that best explains the observed scalp distribution (along with satisfying other implicit or explicit constraints). When I read a paper that reports source localization models, I always wonder what other distributions would fit the data as well as, or almost as well as, the reported solution. Are all reasonable solutions similar to the reported solution? Or are there other solutions that are quite different from the reported solution but fit the data and the constraints almost as well? After all, the presence of noise in the data implies that the correct solution will not actually fit the data perfectly, so a solution that explains only 97 percent of the variance may be closer to the correct solution than a solution that accounts for 100 percent of the variance.

Figure 7.5 shows a simulation presented by Koles (1998) that illustrates this problem. Koles created a distributed source by using several nearby dipoles arranged along a curved surface, and fit an equivalent current dipole to the scalp distribution produced by this distributed source. He placed model dipoles systematically at a variety of locations, and measured the residual error for each of these locations. As the right side of figure 7.5 shows, the error was lowest near the location of the best equivalent current dipole, but there wasn't much difference in error over a fairly wide range of locations. This is exactly the sort of information that one needs to know when trying to evaluate a source localization model.

In my view, it is misguided to attempt to find a unique solution given the uncertainties inherent in ERP localization. A better approach would be to report the entire range of solutions that fit the data and constraints to some criterion level (e.g., a fit of 95 percent or better). Moreover, it would be useful to report the range of

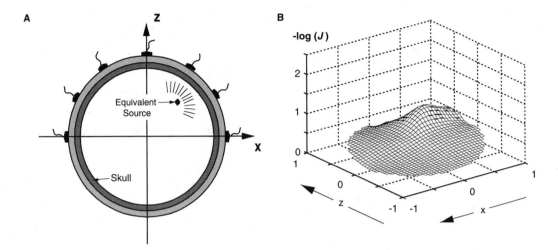

Figure 7.5 Example of a distributed electrical source (A) and the amount of error that would occur in estimating this source by a single dipole over a range of positions (B). Although a single point of minimum error exists, the amount of error varies little over a fairly wide range. (Adapted from Koles, 1998. © 1998 Elsevier Science Ireland Ltd.) Thanks to Zoltan Koles for providing electronic versions of these images.

solutions obtained as one adds and removes various constraints. If, for example, one were to find high levels of estimated activity in a particular region in almost any solution, no matter what constraints were used, then this would give us considerable confidence that this region really contributed to the observed ERPs.

A few investigators have explored this general sort of approach. For example, as described in the section on equivalent source dipole approaches, Huang, Aine, and their colleagues (1998) have developed a multi-start approach in which the dipole localization procedure is run hundreds or thousands of times with different starting positions. This makes it possible to see which dipole locations are found frequently, independent of the starting positions or even the number of dipoles in the model. This approach potentially solves the most significant shortcomings of equivalent source dipole approaches. In particular, the solutions are largely operator-independent, and it is possible to assess the likelihood that a given dipole location occurred because of the starting positions of the dipoles or because of incorrect assumptions about the number of dipoles.

Although the multi-start approach addresses these shortcomings, it still falls short of providing a quantitative description of the probability that a particular brain area contributed to the observed ERP data. That is, a dipole may have been found in a given region in some percentage of the solutions, but the localization approach does not guarantee that the space of adequate solutions is sampled completely and evenly. However, Schmidt, George, and Wood (1999) have developed a distributed source localization technique based on Bayesian inference that provides a more sophisticated means of assessing probabilities. This technique is similar to the multi-start technique insofar as it generates thousands of potential solutions. However, its basis in Bayes's theorem allows it to provide a more complete and quantitative description of the space of possible solutions. This is the most promising localization technique that I have seen. Unfortunately, it has not yet been widely applied to real experiments, and other groups of researchers have not yet thoroughly explored its limitations. Nonetheless, the more general principle embodied by this approach and the multi-start approach—which systematically explore the space of likely solutions—seems like the best direction for the development of source localization techniques.

Recommendations

I will end this chapter by providing some recommendations about whether, when, and how you should use source localization techniques. My basic conclusion is that ERP localization is extremely difficult, and it should be attempted only by experts and only when the solution space can be reduced by well-justified constraints, such as structural MRI data and the combination of electrical and magnetic data. In addition, the most commonly used techniques are useful primarily for obtaining converging evidence rather than providing a conclusive, stand-alone test of a hypothesis, although ongoing developments may someday allow source localization data to provide definitive results.

Source Localization and Scientific Inference

To assess the value of source localization techniques, it is useful to put them into the context of general principles of scientific inference. Perhaps the most commonly cited principle of scientific inference is Popper's (1959) idea of falsification. A commonly used, although less commonly cited, extension of this idea is Platt's (1964) notion of *strong inference*, in which the best experiments are those that differentiate between competing hypotheses, supporting one and falsifying the other.

How do source localization techniques fare when judged by these standards? Not well. I don't think anyone really believes that a single source localization model can conclusively falsify a hypothesis or definitively decide between two competing hypotheses. There are simply too many uncertainties involved in source localization. On the other hand, it is rare that a single experiment using any method is 100 percent conclusive, so this standard may be unrealistic.

A more flexible approach is to apply Bayes's Theorem to scientific inference. In this context, we can summarize Bayes's Theorem by stating that a new result increases the probability that a hypothesis is true to the extent that (a) there is a high probability of that result being true if the hypothesis is true, and (b) there is a low probability of that result being true if the hypothesis is false. In other words, a finding that is consistent with a hypothesis does not give us much more faith in the hypothesis if the finding is likely even if the hypothesis is wrong.

In this context, a given source localization model will have value to the extent that it not only supports a specific hypothesis but is also unlikely to have been obtained if the hypothesis is false. It's the second part of this equation that is especially problematic for source localization models, at least as researchers typically use them. As discussed in the previous section, source localization models provide an estimate of the internal distribution of electrical activity, but they do not typically quantify the probability that the estimate is incorrect (which is related to the probability that the finding would be obtained even if the hypothesis is false). However, this problem can be overcome, at least in principle. For example, the probabilistic approaches described in the previous section are designed to provide information about the range of possible solutions, making it possible to assess the probability that activity would appear in a given location in the models even if the corresponding brain location were not truly active. Thus, although most source localization methods are not well suited for this kind of scientific inference, this does not appear to be an intrinsic limitation of the entire source localization enterprise.

Another commonly cited principle of scientific inference is the idea of *converging evidence*, which was first developed in the context of perception research (Garner, Hake, & Eriksen, 1956) but is now widely used in cognitive neuroscience. The basic idea is that many interesting questions about the mind cannot be answered by means of any single method, but a clear answer can be obtained when many methods with different strengths and weaknesses converge on the same conclusion. This is a common use of source localization models. That is, the researchers understand that the models are not conclusive evidence that the ERPs are generated in specific brain regions, but they believe that the models are valuable insofar as they converge with data from other sources. Box 7.1 provides an example of this from my own research.

Until source localization techniques routinely provide meaningful, quantitative information about the probability that a given model is correct, the main role of source localization models will be to provide converging evidence. However, not all cases of converging evidence are created equal: If a model uses weak methods to create a given source localization model, then this model will provide only weak converging evidence. And the value of such models is questionable, especially given the time and expense often involved in the modeling process. At present, source local-

Box 7.1 Converging Evidence

The following is an example of how I have used source localization to provide converging evidence for a specific hypothesis. My initial visual search experiments examining the N2pc component suggested that this component reflects the focusing of attention onto a target and filtering out irrelevant information from the distractor objects (Luck & Hillyard, 1994a, 1994b). This seemed similar to the types of attention effects that Moran and Desimone (1985) observed in single-unit recordings from area V4 and from inferotemporal cortex, but I had no way of localizing the N2pc to these areas. Then Leonardo Chelazzi conducted a series of follow-up studies in Desimone's lab using visual search tasks that were more similar to the tasks that I had used to study the N2pc component (Chelazzi et al., 1993, 1998, 2001). The onset of the attention effects in these single-unit studies was remarkably similar to the onset time of the N2pc component, and this suggested that the N2pc component might reflect the same neural activity as the single-unit attention effects. To test this hypothesis, I conducted a series of N2pc experiments that paralleled Chelazzi's single-unit experiments, and I found that the N2pc component responded to several experimental manipulations in the same way as the single-unit attention effects. To provide converging evidence, I collaborated with Max Hopf and Hajo Heinze on a combined ERP/ERMF study of the N2pc component using the cortically constrained minimum norm approach (Hopf et al., 2000). The resulting source localization model was consistent with a source in the general area of the human homologues of monkey V4 and IT (with an additional source in posterior parietal cortex). In this manner, the source localization data provided converging evidence for a link between the N2pc component and a specific type of neural activity (see Luck, 1999 for an extended discussion of this general approach, which combines traditional hypothesis testing with source localization).

ization models provide reasonably strong converging evidence only if they are the result of state-of-the-art methods and only if they are developed thoughtfully and carefully.

Specific Recommendations

My first specific recommendation is to avoid techniques that involve substantial input from the operator (which is true of many, but not all, equivalent current dipole approaches). These techniques are so prone to experimenter bias that they can provide only the weakest sort of converging evidence. In fact, I would argue that these models are often worse than no models at all, because they provide the illusion of strong evidence when in fact the evidence is weak. The one exception to this recommendation is that these approaches may be adequate when a combination of three criteria are met: (1) the data are very clean; (2) you can be sure that only one or perhaps two dipoles are present; and (3) you have good reason to believe that the electrical activity is relatively focused rather than being distributed over a large region. Localization is fairly easy under such conditions, and validation studies have shown that localization errors average approximately 1 cm or less when these criteria are met (see, e.g., Cuffin et al., 1991; Leahy et al., 1998).

My second specific recommendation is to obtain structural MRI scans from each subject so that you can create a reasonably accurate head model and use one of the cortically constrained approaches (which typically involve distributed source solutions rather than equivalent current dipole solutions). For most experiments in cognitive neuroscience, it is very likely that the activity is generated exclusively in the cortex with a perpendicular orientation, and this provides a powerful constraint that reduces the solution space considerably. The most common versions are the depth-weighted minimum norm and LORETA techniques, described previously in this chapter. LORETA is well suited for situations in which (a) you want to determine the center of each activated region, (b) you do not care about the spatial extent of the activated regions, and (c) the activated regions are well separated from each other. If these conditions are met, LORETA appears to work quite well (see, e.g., the impressive LORETA/fMRI correspondence obtained by Vitacco et al., 2002). If these conditions are not met, I would recommend using the depth-weighted minimum norm approach.

My third recommendation is to use difference waves to isolate a single component (or small set of components; see chapter 2 for more discussion). The more components are active, the more of a mess you will have to sort out. Equivalent current dipole approaches become particularly problematic when more than a few sources are present, but this can also be a problem for distributed source approaches.

My fourth recommendation is to record ERMFs in addition to ERPs. ERMFs have two major advantages over ERPs. First, they are not blurred and distorted by the high resistance of the skull, leading to greater resolution and a smaller space of possible solutions (see, e.g., the simulation results of Leahy et al., 1998). Second, because biological tissues are transparent to magnetism, it is not necessary to create a model of the conductivities of the brain, skull, and scalp, and this eliminates one possible source of error. ERMF recordings do have a couple disadvantages, though. First, they are very expensive, both in terms of the initial capital investment and the maintenance costs (particularly the coolant). Second, ERMF recordings will not be able to detect sources that are deep or perpendicular to the surface of the head (note, however, that this becomes an advantage rather than a disadvantage when ERMFs are combined with ERPs). In my experience, source localization is just too uncertain when based on ERPs alone.

The bottom line is that source localization is extremely difficult, and any serious attempt at localization will require sophisticated methods and considerable costs (both in terms of time and money). If you simply record ERPs from a large number of channels and try to fit a half dozen dipoles to the data with no additional constraints, it's not clear what you will have learned. At best, you will gain some weak converging evidence. At worst, you will be misled into believing in a solution that is simply incorrect.

I would like to end by noting that the clear strength of the ERP technique is its temporal resolution, not its ability to localize brain function. It is therefore sensible to use this technique primarily to answer questions that require temporal resolution, leaving questions about localization of function to other techniques. When a question requires a combination of temporal and spatial resolution, a combination of ERPs, ERMFs, structural MRI scans, and fMRI data may provide reasonably strong evidence, but the commonly used methods for localizing ERPs/ERMFs do not make it clear how strong the evidence is. As new techniques are developed— particularly those based on probabilistic approaches—we may eventually get to the point where we can have a reasonably high (and known) level of certainty.

Suggestions for Further Reading

The following is a list of journal articles and book chapters that provide useful information about ERP and MEG source localization.

Dale, A. M., & Sereno, M. I. (1993). Improved localization of cortical activity by combining EEG and MEG with MRI cortical surface reconstruction: A linear approach. *Journal of Cognitive Neuroscience, 5,* 162–176.

George, J. S., Aine, C. J., Mosher, J. C., Schmidt, D. M., Ranken, D. M., Schlitt, H. A., Wood, C. C., Lewine, J. D., Sanders, J. A., & Belliveau, J. W. (1995). Mapping function in the human brain with magnetoencephalography, anatomical magnetic resonance imaging, and functional magnetic resonance imaging. *Journal of Clinical Neurophysiology, 12,* 406–431.

Hämäläinen, M. S., Hari, R., Ilmonieni, R. J., Knuutila, J., & Lounasmaa, O. V. (1993). Magnetoencephalography—theory, instrumentation, and applications to noninvasive studies of the working human brain. *Review of Modern Physics, 65,* 413–497.

Koles, Z. J. (1998). Trends in EEG source localization. *Electroencephalography & Clinical Neurophysiology, 106,* 127–137.

Luck, S. J. (1999). Direct and indirect integration of event-related potentials, functional magnetic resonance images, and single-unit recordings. *Human Brain Mapping, 8,* 15–120.

Mangun, G. R., Hopfinger, J. B., & Jha, A. P. (2000). Integrating electrophysiology and neuroimaging in the study of human brain function. In P. Williamson, A. M. Siegel, D. W. Roberts, V. M. Thandi & M. S. Gazzaniga (Eds.), *Advances in Neurology* (Vol. 84, pp. 33–50). Philadelphia: Lippincott, Williams, & Wilkins.

Miltner, W., Braun, C., Johnson, R., Jr., Simpson, G. V., & Ruchkin, D. S. (1994). A test of brain electrical source analysis (BESA): A simulation study. *Electroencephalography & Clinical Neurophysiology, 91,* 295–310.

Pascual-Marqui, R. D., Esslen, M., Kochi, K., & Lehmann, D. (2002). Functional imaging with low-resolution brain electromagnetic tomography (LORETA): a review. *Methods & Findings in Experimental & Clinical Pharmacology, 24 Suppl C,* 91–95.

Scherg, M., Vajsar, J., & Picton, T. (1989). A source analysis of the human auditory evoked potentials. *Journal of Cognitive Neuroscience, 1,* 336–355.

Schmidt, D. M., George, J. S., & Wood, C. C. (1999). Bayesian inference applied to the electromagnetic inverse problem. *Human Brain Mapping, 7,* 195–212.

Snyder, A. (1991). Dipole source localization in the study of EP generators: A critique. *Electroencephalography & Clinical Neurophysiology, 80,* 321–325.

This chapter describes how to set up an ERP lab, including advice on selecting equipment and software for stimulus presentation, data collection, and data analysis. There are certainly many ways to set up an ERP lab, and some factors will depend on the topic area of your research. However, the suggestions in this chapter will provide a good starting point for almost anyone who is setting up an ERP lab for the first time. Also, many of the suggestions in this chapter are based on my experience, but there are often other good ways of achieving the same goals, and it is worth asking a variety of ERP researchers how they set up their labs. If you already have access to an ERP lab, you may find some good ideas for improving the lab in this chapter (see especially box 8.1 later in this chapter).

The Data Acquisition System

Computers

Figure 8.1 shows a diagram of a generic ERP data acquisition system. It is usually desirable to have at least two and possibly three computers. One computer presents the stimuli, and a second computer records the EEG. These functions are not ordinarily combined into a single computer, because each requires precise timing, and it is difficult to coordinate two real-time streams of events in a single computer. It is also very useful to have a third computer that provides real-time information about the subject's performance on the task. The stimulus presentation computer sometimes provides this

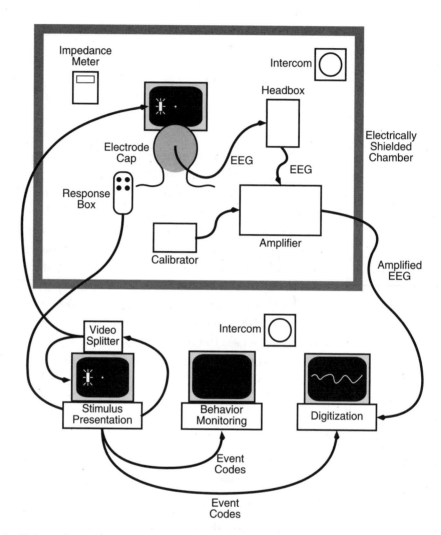

Figure 8.1 Major components of a typical ERP recording system.

function, but in this case performance information is usually summarized at the end of each trial block rather than being displayed in real time (and a real-time display is definitely better).

The computers will need some means of communicating with each other in real time so that *event codes* can be sent to the digitization computer whenever an event of some sort occurs (e.g., a stimulus or a response). These event codes are used as the time locking points for averaging, so the timing must be precise. A consistent delay is not a big problem, because you can shift the averages in time to compensate for the delay. A variable delay, however, is usually difficult to compensate for, and it will have the effect of smearing out the ERP waveforms in time, distorting the onset and offset times of the components and experimental effects (just like a low-pass filter).

The stimulus presentation computer's video output is usually directed into a video splitter that serves two functions. First, it splits the video signal so that the experimenter can see what the subject is seeing (this is essential so that the experimenter can intelligently monitor the EEG display and detect problems, such as blinks and eye movements that certain stimuli may trigger). Second, it amplifies the video signal so that one can use a relatively long cable to get the video signal into the recording chamber. It is important to use a high-quality, shielded video cable from the splitter into the chamber to minimize electrical noise inside the chamber and to avoid degradation of the video signal. As discussed in chapter 3, it's a good idea to enclose the video monitor in a Faraday cage to avoid bringing a large noise signal into the recording chamber. I provide more information about the video system later in this chapter, in the section on stimulus presentation.

Seating

In experiments using visual stimuli, the subject is seated at a specific distance from the video monitor. Even with a Faraday cage, some electrical noise may be picked up from the monitor, so the subject should not be too close to it. I would suggest a minimum distance of 70 cm, and 1–2 m is even better if the chamber is large enough.

It is very important for the subject to be seated in a comfortable position. In the early days of ERP recordings, very few electrodes were used and auditory stimuli were more common than visual stimuli. A large, padded recliner was the most sensible chair for such experiments. This kept the subject comfortable, and minimized any need for subjects to use their neck muscles to support their heads, which in turn minimized EMG noise (see chapter 4). However, when electrodes are placed over the back of the head, a recliner doesn't work as well: the head puts pressure on the electrodes, and any small head movement will cause the electrodes to move on the scalp, producing a large and sudden voltage shift.

If your experiments will require recording from electrodes on the back of the head, I would not recommend using a recliner. Instead, I would recommend a high-quality office chair that provides good lumbar support and has an easy mechanism for height adjustment. It is also preferable to use a chair that is on glides rather than casters so that subjects do not move the chair around inside the chamber. The position of the chair on the floor should be marked so that you can correctly reposition the chair if it gets moved. I would also recommend against using any kind of chin rest or other head stabilization. Although you might think that this would reduce strain on the neck muscles, my experience has been that chin rests become uncomfortable after 15 minutes or so (and most ERP experiments require at least an hour of recording time).

Behavioral Responses

Most cognitive ERP experiments use a small set of behavioral response alternatives, and only two to four buttons are usually necessary. The most convenient device for responses is usually a small, lightweight, hand-held device of some sort. My lab uses computer video game controllers. It is easy to find game controllers that are very comfortable and lightweight, and they usually offer at least four response buttons (which is enough for most experiments). Various custom response devices are also possible, but they are not usually as easy to hold and use as a game controller. A computer keyboard is not usually very good, because it is too big to rest on the subject's lap, and if it's on the table in front of the subjects, EMG noise is likely to occur as the subjects hold their arms up to the keyboard. If a keyboard is necessary because of a large number of response alternatives, the data around the time of the response may be contaminated by EMG noise, and eye movement artifacts may be seen if the subject needs to look at the keyboard to find the right key to press. However, it may be possible to design the experiment so that the data around the time of the response are not important. Also, it may be possible to use a keyboard tray to minimize the need to hold the hands upward and outward to reach the keyboard.

Devices such as keyboards and game pads are not designed for real-time electrophysiological recording, and a delay may occur between the time of the buttonpress and the time that the computer records the buttonpress. Response times are usually so variable that this small error will not matter. However, it may be a significant problem if you are doing response-locked averaging. In such cases, you will probably need some sort of custom response device with known and consistent temporal properties. One way to ensure precise timing is to use a response device with an analog output that can be recorded via the ADC along with the EEG.

Electrode-Amplifier-Computer Connections

The EEG is picked up by the electrodes and travels through cables to the amplifier system. Before they are amplified, the signals are miniscule, and any electrical noise that the cables pick up will be relatively large compared to the EEG. Consequently, the cables between the electrodes and the amplifier system should be kept as short as is practical. It is now possible to purchase electrodes with built-in pre-amplifiers, which will increase the size of the EEG signal relative to any noise picked up by the cables. This is a good idea conceptually, but I have not tested these electrodes and I don't know how well they work in practice.

Many EEG amplifier systems have a *headbox* that provides a flexible means of connecting the electrodes to the amplifier. Flexibility is important. For example, you want to be able to select a common reference electrode for most of the channels (e.g., a mastoid or earlobe reference) while allowing other channels to use a different reference electrode (e.g., eye movements are usually recorded as the voltage between electrodes on either side of the eyes). Because the output of the headbox is not amplified, the cable from the headbox to the amplifier should be relatively short. Optimally this cable should be shielded, especially if the actual amplifier is outside the recording chamber. If the amplifier is inside the chamber, an unshielded ribbon cable will probably be sufficient (a ribbon cable keeps the active, reference, and ground signals close to each other so that they pick up the same noise signals, and this allows the differential amplifier to subtract away the noise).

If the amplifier is connected to AC power, it is usually best to place it outside the recording chamber to avoid inducing noise in the electrodes. If it is battery powered, it can probably be placed inside the chamber, which is good because only amplified signals will leave the chamber, where electrical noise is much greater. Once the signals have been amplified, they are usually much larger than any induced electrical noise, so you may use longer cables to connect the amplifier to the digitization computer. However, some noise may still be picked up at this stage, and some shielding may be necessary even after the signals have been amplified. In one of the recording rooms in my lab, we found that shielded cables were necessary to avoid picking up noise between the amplifier and the digitization computer.

Recording Chamber

One of the most expensive elements of an ERP lab is the electrically isolated chamber. These chambers are usually designed as sound-attenuating chambers for hearing testing, broadcasting, or auditory research, but some manufacturers offer electrical shielding options. I have chambers made by Industrial Acoustics Company and Acoustic Systems, and they work quite well. In the late 1990s, I paid approximately US $11,000 for a medium-sized chamber (including installation costs).

Is an electrically isolated chamber really necessary? Not always. If there are not many significant sources of electrical noise nearby, you might be able to get away with using a low-pass filter with a half-amplitude cutoff around 30 Hz to get rid of line-frequency noise. If you are just starting to do ERP research and you are not sure how many experiments you will be doing, this might be a reasonable choice. However, if you plan to do ERP research for several years, the cost of a chamber isn't really that high, and it's a worthwhile investment. Remember that any significant decreases in signal-to-noise ratio will be very costly in terms of the number of trials (or subjects) needed to get reliable ERP waveforms (see chapter 3).

Don't assume that the chamber effectively eliminates all sources of electrical noise. Chambers often include AC outlets, fans, and lights that may cause substantial electrical noise. In fact, the first chamber I ever used induced so much noise that I was able to get cleaner recordings outside the chamber than inside. I eventually found the circuit breaker for the chamber, and turning it off eliminated the noise. You can use the device described in chapter 3 (see figure 3.3B) to find sources of noise inside the chamber.

Accessories

A few accessories are important in an ERP lab. First, it is usually necessary to have an impedance meter to test electrode impedances while attaching the electrodes. Some amplification systems have this function built into them, which is very useful for determining if impedance problems are occurring once you've started collecting data. Second, some sort of calibrator is necessary for measuring the actual gain of the amplifiers. Many amplification systems have a built-in calibrator, but you can purchase an external calibrator if your amplification system does not have its own calibrator. If possible, the calibrator should be able to produce square-wave pulses triggered by the stimulus presentation computer. This will help you to determine if there are any time delays in your event codes, and it will also make it easy for you to see exactly what kinds of distortion the amplifier's filters produce.

A third important accessory is an intercom system that will allow you to communicate with the subject during the recordings. When purchasing an intercom system, keep in mind that you don't want a system that is AC powered, because this will introduce electrical noise into the recording chamber. I've tried several different types of systems, and my preferred intercom system consists of a pair of microphones and powered speakers (which you can purchase at many places, such as Radio Shack). One microphone is attached to the ceiling of the recording chamber and is connected to a powered speaker outside the chamber. The other microphone is mounted near the experimenter and is connected to a powered speaker inside the chamber. The speaker inside the chamber should be powered by a battery. The advantages of this system over a conventional intercom are that a) it allows simultaneous two-way communication between the subject and experimenter (which is more natural than pressing a button when you want to talk), and b) the fidelity is quite a bit higher.

Finally, if you will be presenting visual stimuli, I strongly recommend that you enclose the video monitor in a Faraday cage to reduce electrical noise, as described in chapter 3 (see figure 3.3A). Some additional suggestions for recording high-quality data appear in box 8.1.

Box 8.1   Keeping Subjects Happy

ERP experiments tend to be long and boring, with trial after trial of the same basic task. To ensure that you are collecting the highest quality data possible, it is important to keep your subjects happy and relaxed. If they are unmotivated or become bored, they may not pay close attention to their performance, weakening your effects. Moreover, bored subjects tend to become tense and uncomfortable, leading to muscle noise and movement artifacts. By keeping your subjects happy and relaxed, you will get larger and more consistent effects, which will save you lots of time in the long run. Here are four suggestions for accomplishing this.

First, talk to the subject. As you are applying the electrodes, chat with the subject, asking about school, work, hobbies, family, sports, and so on (but stay away from potentially inflammatory topics such as politics). Make sure that the conversation revolves around the subject rather than yourself. By doing this, you will ingratiate yourself to the subject, and the subject will be more likely to try to please you during the experiment. In addition, when there are breaks in the experiment, you should continue to converse with the subject to keep the subject alert and interested. Some subjects just want to finish the experiment as rapidly as possible, and you should respect this. But if the subject is at all interested in talking, you should do so. This is also a good time to provide some positive feedback to the subject about behavioral performance, blinks, and eye movements, as well as encouraging the subject to improve if necessary. If a subject isn't doing well (e.g., in terms of behavioral accuracy or artifacts), don't be shy about telling them that, because poor performance and artifact-laden data won't be of much use to you. You should also make sure that the subject understands exactly what is going to happen and when (especially during the electrode application process). People are much more relaxed when they know what to expect.

Second, make sure that the blocks of trials are a reasonable length. If the blocks are too long, the subject's attention is likely to wane toward the end. I find that trial blocks of 5–7 minutes, separated by 1–2 minutes of rest, are optimal in most experiments. In addition, you will usually ask the subject to suppress blinks and eye movements during the recordings, and it is difficult to do this for a long period of time. Some experiments consist of brief trials separated by an intertrial interval of 1–3 seconds, and this provides a time for subjects to blink (although the offset of the blink during the intertrial may contaminate the early portion of the ERP on the next trial). If this is not possible, I would recommend providing short breaks of 15–20 seconds after every 1–2 minutes.

Third, due to the long duration of a typical ERP experiment, it is helpful to provide snacks and drinks, usually 30–50 percent of the way through the

Box 8.1 (continued)

session. Drinks including caffeine can be particularly useful in helping subjects to maintain alertness (although this may be contraindicated in some
experiments). This helps to keep the subjects awake, alert, and friendly.

My fourth recommendation is something that I started doing when I first
set up my own ERP lab, and I don't know if anyone else does it. Specifically,
we play background music in the chamber while the subject is doing the task.
In fact, we suggest to the subjects that they bring CDs of their favorite music
with them (and the musical genres have included classical, pop, rock, metal,
rap, country, electronic/ambient, and just about everything else imaginable).
Of course, the music produces some distraction from the task, and the
sounds will generate ERP activity. However, I believe that the music is less
distracting than the alternative, which is mind-numbing boredom. And any
ERP activity generated by the music will be unrelated to the stimuli and will
just add a small amount of noise to the EEG. I suspect that any additional
EEG noise created by the music is more than balanced by a reduction in other
forms of EEG noise, such as alpha waves. Of course, there are situations in
which background music is a bad idea. For example, it would be problematic
in many studies using auditory stimuli, and it might cause too much distraction in some types of subjects. But if these circumstances don't apply to you,
I would definitely recommend playing background music. I've never done a
direct comparison of data collected with versus without background music,
but I strongly suspect that background music leads to a moderate increase in
the overall quality of the data. If you play music, make sure that you use
shielded speakers and that the cables leading from the stereo to the speakers
are shielded. Speaker cables are not usually shielded, but you can easily en-
case them in some form of shielding, such as Wiremold.

Choosing Electrodes, Amplifiers, and Digitization Software

New ERP users occasionally ask for my advice about what electrodes, amplifiers, and software to buy for recording ERPs; this section summarizes my usual advice. Given that I haven't thoroughly
tested all of the commercially available systems, it would be inappropriate for me to recommend specific manufacturers. However, I can provide some general advice about how to go about
selecting data acquisition equipment.

First, however, I'd like to provide a caveat. The suggestions here
are my own opinions, and they may be different from the opinions

of other researchers (and especially certain equipment manufacturers). Consequently, I would recommend seeking advice from a variety of experienced ERP researchers (but not from the manufacturers, who are quite naturally biased).

Electrodes My electrode recommendation is simple: Old-fashioned, low-tech electrode caps are the best option for the vast majority of new ERP users. As discussed in chapter 3, some newer systems allow you to apply a large number of electrodes in a short amount of time, whereas traditional electrodes require you to abrade the skin under each electrode, which takes quite a bit of time when applying large numbers of electrodes (e.g., an hour for sixty-four electrodes compared to 15 minutes for sixteen electrodes). However, these systems tend to be more expensive than conventional systems, and the high electrode impedances in these systems can lead to a substantial increase in low-frequency noise from skin potentials. Moreover, large numbers of electrodes do not provide a very significant advantage in most ERP studies, but they can provide a disadvantage because there are more opportunities for electrodes to misbehave as the number of electrodes increases. Consequently, I would recommend standard electrode caps with between twenty and thirty-two electrodes for most ERP studies.

Fast-application, high-impedance electrode systems may seem particularly appealing to researchers who study subjects who do not easily tolerate being poked and prodded, such as infants and young children. However, some researchers have been recording ERPs from such subjects for many years with conventional electrodes, so it can be done (although it requires good measures of both ingenuity and patience). Moreover, given that it is difficult to collect a large number of trials from such subjects, their data are far too noisy to be suitable for source localization procedures, so there is little or no advantage to having large numbers of electrodes. Thus, I would recommend using conventional electrodes even for infants and young children.

Amplifiers When selecting an EEG amplifier, there are a few key specifications that you should examine and several features that you may wish to have (depending on your budget). The most important specifications are the input impedance and the common mode rejection (see chapter 3 for definitions of these terms). The input impedance should be at least 100 KΩ, and the common mode rejection should be at least 100 dB. The essential features are (a) the ability to select a common reference for many sites but separate references for other sites, and (b) a wide range of filter settings. In particular, I would recommend having a low-pass filter that can be set to half-amplitude cutoffs of 20–40 Hz, 80–120 Hz, and 300–700 Hz, and a high-pass filter that can be set to cutoffs of approximately 0.01 Hz (for most recordings) and approximately 0.1 Hz (for use with especially troublesome subjects). In some amplifiers, the lowest half-amplitude cutoff is 0.05 Hz, which is five times higher than the 0.01 Hz that I would recommend; this is right at the border of what I would consider acceptable. It would also be worth asking the manufacturer to provide the impulse response function of the filters so that you can assess the time-domain distortions that the filters will produce (if they can't provide the impulse response functions, then that should be a warning sign about the manufacturer). I would recommend against DC amplifiers that don't have a high-pass filter, unless you plan to record very slow voltage shifts.

Here are some features that are useful, but not absolutely necessary:

1. Multiple gain settings. It can be useful to increase or decrease the gain for testing purposes or to match the input range of the analog-to-digital converter.

2. Impedance checking. It's convenient to be able to test impedances without disconnecting the subject from the amplifier. It's even better if the system can automatically warn you if a channel's impedance exceeds some user-defined level.

3. Calibration. It's useful for the amplifier to have a built-in calibrator.

4. Notch filter. Although notch filters are generally a bad idea, they are sometimes a necessary evil (see chapter 5).

5. Independent or linked filter settings. You will usually want to use the same filter settings for every channel, but it's occasionally useful to have different filter settings for a few of the channels.

6. Computer-controlled settings. If a computer can set the amplifier's settings on the basis of a user-defined profile, this will decrease the probability that another user will unintentionally change the settings.

Digitization Software   Once you have electrodes and an amplifier, you will need a computer and software for digitizing the EEG. In some cases, the electrodes will dictate the choice of amplifier, and the amplifier may have an associated digitization program. Most digitization programs have the basic features that you will need, so you should choose whatever seems convenient and will work with the other components of your system. The essential features of a digitization program (and associated analog-to-digital converter) are as follows:

7. The analog-to-digital converter must have at least twelve bits of precision. Many now offer sixteen bits of precision, but given the poor signal-to-noise ratio of the EEG signal, this is not a very big advantage. A sixteen-bit converter allows you to use a lower amplifier gain, reducing the probability that the amplifier will saturate. However, saturation is usually caused by large, slow voltage shifts that will contaminate your data even if the amplifier does not saturate.

8. There must be a convenient means of sending event codes to the system. Some systems merely use one channel of the analog-to-digital converter to encode a pulse whenever an event occurs, and the actual nature of the event is stored on another computer (usually the stimulus presentation computer). This makes averaging the stimuli less convenient, so it is better for the digitization computer to receive codes indicating the precise nature of each

stimulus. This is usually done with eight-bit digital codes (it is convenient to have even more than eight bits for some experiments).

9.   The system must allow continuous EEG recording. Many ERP paradigms involve stimuli that are separated by long periods of time with no ERP-eliciting events, and it is therefore possible to record just the EEG segments surrounding the stimuli, pausing during intertrial intervals. This is called *epoch-based recording*. However, you may later wish that you had recorded a longer epoch. For example, you may find an effect that is still growing in amplitude at the end of the epoch, or you may want to evaluate prestimulus overlap more thoroughly. Epoch-based recording does not allow this. The alternative (*continuous recording*) is to record the EEG continuously and extract whatever epochs are necessary during the averaging process. All of the data are saved, so it is possible to re-average the data with a different epoch at a later time. This requires more storage space, but storage space is so cheap now that it's not a problem to record continuously. Thus, you will want a system that can do continuous recording. The option to do epoch-based recording may sound attractive, but I would recommend against epoch-based recording because of the loss of flexibility.

10.   The system must allow a convenient way to view the EEG in real time as it is being recorded. Real-time monitoring is essential so that the experimenter can identify problems with the electrodes, high levels of artifacts (especially ocular artifacts and muscle activity), and subject weariness as indicated by high levels of alpha activity (weariness can be a significant issue because ERP experiments usually last several hours). It is essential to be able to adjust the vertical scale of the signal to match the size of the EEG signal, and it is also useful to be able to adjust the horizontal (time) scale. The system should also include some means of seeing the arrival of event codes. It is very convenient if the event codes appear in a position that is synchronized to the EEG display so that you can view the relationship between event codes and artifacts.

Some additional options that may be useful are as follows:

11. The ability to view the EEG in the frequency-domain as well as the time domain. This makes it easier to determine the level of various noise signals, such as line-frequency electrical noise, and to notice when alpha levels are getting high (which is often a sign of drowsiness).

12. The ability to save the data in multiple formats, including text files. Text files are huge, but they have the advantage of being easy to convert into other formats.

13. The ability to do channel mapping, in which the amplifier channels are sampled in an arbitrary, user-controlled order. This makes it easy to put the data into a useful order, skipping channels that are not necessary for a given experiment.

14. The ability for the user to define the spatial layout of the EEG display. This may make it easier to see patterns in the EEG that are indicative of artifacts. Some systems can even display maps of EEG frequency bands in real time.

15. The ability to do simple on-line averaging. In some experiments, for example, subjects may tend to make small but systematic eye movements that are difficult to see on individual trials but can be easily seen in averaged waveforms. By doing on-line averaging, it may be possible to identify and correct artifacts such as this during the recording session rather than throwing out the subject's data afterwards.

Basic digitization software is fairly straightforward, so you don't need to worry too much about getting software that will work well. The only tricky aspect of digitizing the EEG is that both the EEG and the event codes must be acquired in real time, with no temporal errors. Modern computer operating systems make this difficult, so I would not recommend that you write your own digitization software unless you have considerable expertise with real-time programming.

Once you have purchased a digitization system, it is very important that you test it very carefully. The most common problem is that delays will be introduced between stimuli and the event codes stored on the digitization computer. In fact, my lab recently found that one of the most common commercial digitization systems introduced delays that increased gradually over the course of a trial block. When we contacted the manufacturer, they told us that this problem arises when their software is used with a particular brand of computer, and the problem disappeared when we switched to a different computer. But we never would have noticed the problem if we hadn't tested the system extensively. You cannot simply assume that your software works correctly.

To test a digitization system, you need to be able to record a reference signal that is triggered by your stimulus presentation system. The easiest way to do this is to use a square-wave calibration signal with an external trigger input (if your amplifier doesn't have such a calibrator, you can buy one). You can trigger the calibrator with your stimulus presentation software and see if the calibration pulse occurs at the same time as the event codes in the EEG data file. The square wave should start at the same time as the event code, and you should test this over a period of many minutes to convince yourself that the timing does not drift over time. Depending on the nature of your system, you may occasionally see an offset of one sample period; this can occur occasionally even if the timing only off by a microsecond or two. But if this happens on more than 10 percent of event codes, or if the errors are more than one sample period, you should contact the manufacturer.

The Data Analysis System

My lab uses a terrific package of custom ERP analysis software, but this software is not available to the general public. Thus, my advice in this section is based on what I would do if I did not have access to this system.

I have taken a look at some of the commercial ERP analysis systems that are available, and I haven't been favorably impressed by them. They are full of bells and whistles, but they are expensive and they don't seem to do exactly what I need them to do. One reason for this is that scientific research is generally focused on doing things that have never been done before or that need to be done in a way that depends on the details of the research domain. This makes it hard for software companies to write ERP analysis software that suits the specific needs of the diverse set of researchers and can anticipate future needs. They also try to make the software easy to use, but ease of use is usually inversely related to power and flexibility.

Instead of buying commercial software, you may want to write your own ERP analysis software. I've done a fair amount of this over the years, and it's very time consuming. Flexible data analysis programs that can be used in many experiments takes a very long time to write, especially in general-purpose programming languages such as C, Pascal, and BASIC. Special-purpose programs written for a single experiment take less time to write, but your research will go slowly if you have to write new programs for the analysis of every experiment.

Perhaps the best compromise is to write your own software, but to do it in a development system that is designed for numerical processing, such as MATLAB. MATLAB is designed for performing mathematical computations on arrays and matrices of numbers, and this is exactly what ERP waveforms are. That is, in a typical experiment, the averaged data can be described as a time $\times$ electrode $\times$ condition $\times$ subject matrix of voltage values. MATLAB contains built-in routines for quickly performing mathematical operations on matrices, and these routines are accessed from a relatively simple but powerful programming language.

To make data analysis with MATLAB even easier, two groups of researchers have developed free, public-domain MATLAB libraries for analyzing EEG/MEG and ERP/ERMF data. One of these packages is called EEGLAB (Delorme & Makeig, 2004, ⟨http://sccn .ucsd.edu/eeglab/⟩). This package includes routines for basic functions, such as artifact rejection and averaging, and it also includes excellent support for some advanced functions, particularly frequency-based analyses and independent components analysis. And it can import EEG data files from a variety of data acquisition systems. The second package is called BrainStorm (Baillet et al., 1999, ⟨http://neuroimage.usc.edu/brainstorm/⟩). It focuses primarily on source localization techniques.

Both of these packages provide a graphical user interface, which is very useful for new users. However, it is also possible to type written commands, and this makes it possible to generate scripts, which are extremely valuable for experienced users. Moreover, once you learn a little bit of MATLAB, you can add your own functions. I would definitely take this approach if I were setting up an ERP lab and didn't have access to my current data analysis package.

Assuming that these free software packages will not fill 100 percent of your needs, this approach requires that you are (a) reasonably competent at computer programming, (b) willing to spend a considerable amount of time learning computer programming, or (c) willing to hire someone else to do some computer programming. If you are a graduate student and you are planning to do ERP research extensively in graduate school and beyond, it would be worth learning some programming. If you are already an advanced researcher and you are planning to augment your research with a modest number of ERP experiments, you probably won't have time to learn to program and write the programs you need.

If you can't do the programming yourself, and you can't pay someone else to do it for you, then you'll have to buy one of the commercial ERP analysis systems. I can't tell you what package to buy, but I can make a few suggestions about some features that are important:

16. Easy methods for averaging the data on the basis of sophisticated criteria, such as sequences of stimuli, correct versus incorrect responses, and responses with various different reaction times. Response-locked averages must be possible.

17. A broad set of artifact rejection functions, such as those described in chapter 4. You should be able to set the parameters on the basis of visual inspection of the EEG. It is also useful to be able to do artifact rejection manually on the basis of visual inspection.

18. The ability to filter the raw EEG and the averaged ERP waveforms with a variety of half-amplitude cutoffs. The impulse response functions of the filters must be clearly described, and the inclusion of filters with gaussian impulse response functions is highly desirable (see chapter 5).

19. A broad set of component measurement routines should be available (see chapter 6).

20. The ability to plot topographic maps showing the distribution of voltage (or current density) over the scalp, preferably using the spherical spline interpolation algorithm (see Perrin et al., 1989).

21. It must be easy to perform mathematical operations on the ERP waveforms, including re-referencing the data and forming difference waves.

22. The ability to import and export data in different formats, particularly text files, is very useful.

23. The ability to automate processing can be very useful. For example, you may find that you need to re-average the data from each subject in an experiment with a longer time epoch, and you will save a lot of time and effort if you can automate this process.

24. Convenient statistical analyses are good to have. You can use standard, general-purpose statistical packages instead, but it is useful to be able to automate the process of measuring and analyzing the data, which is difficult with most general-purpose packages.

25. A convenient but flexible means of plotting the ERP waveforms is essential. You must be able to specify the line types and line colors used for overlapping waveforms, the placement of these sets of overlapping waveforms, and the formatting of the time and voltage axes (see the section on plotting at the beginning of chapter 6). In addition, the system must be able to save the plots in a vector-based file format that can be imported by standard vector-based

graphics packages (bitmapped outputs are not usually suitable for creating figures of ERP waveforms for publication).

The Stimulus Presentation System

Timing of Event Codes

In general, the key to good ERP stimulus presentation is precise timing of the event codes with respect to the stimuli. Modern computer operating systems are not designed to operate in real time, and programs are often interrupted briefly so that the operating system can conduct various housekeeping tasks. These interrupts are not usually noticeable during normal computer usage, but they can introduce timing errors that are large relative to the time scale of ERP recordings (as much as 200 ms). Real-time programming and precise timing was actually easier to achieve with more primitive systems, such as MS-DOS and the Apple II's operating system. Consequently, you should not attempt to create your own stimulus presentation software unless you really know what you are doing (or have access to software development systems that simplify real-time control).

In addition, you should not assume that commercially available stimulus presentation software has the level of precision necessary for ERP recordings. About 15 years ago, I tested one of the most popular stimulus presentation programs of the time, and I found very significant timing problems (it was truly shocking!). Thus, whatever system you use, you should verify the precision of its timing (I'll explain how to do this in the next paragraph). And keep in mind that precision is the key, not accuracy (as these terms are technically defined). That is, a constant, known delay between the stimulus and the event code is fine, because it can be subtracted away by shifting the averaged waveforms by the delay factor. An unpredictably varying delay is not acceptable, however, because there is typically no way to compensate for it.

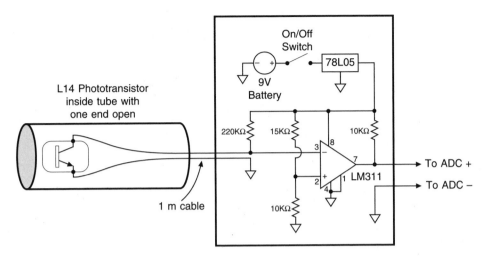

Figure 8.2   A circuit that can be used to measure the light emitted by a small region of a video display. The output can be connected directly to the analog-to-digital converter of your EEG digitization system (assuming that the voltage range is correct). Thanks to Lloyd Frei for designing this circuit.

The best way to test the timing of a stimulus presentation system is to somehow record the stimulus with your data acquisition system. With auditory stimuli, for example, you can place a microphone in front of the speakers and record the microphone's output as if it were the EEG. You can then use your data analysis software to see if the onset of the stimulus occurs exactly at the time of the condition code. This is a bit more difficult with visual stimuli, but figure 8.2 shows a circuit for a simple device that can be used to measure the light being generated by a video monitor. You can point this device toward the part of the screen where the stimuli appear and record the output on your data acquisition system. When you do this kind of testing, I would recommend using a higher-than-normal sampling rate to record the signals (1000 Hz or greater). This will make it easier to see these signals, which may contain very fast transitions. In addition, you should record a large number of stimuli so that you can test for the presence of occasional large timing errors. If you find timing errors, you may be able to eliminate them by disabling any unnecessary hardware and software systems on the computer. For example, you can make sure that no other programs or operating system extensions are running in parallel on the computer, and you can disconnect the computer from networks, unnecessary hard drives, printers, and so on.

Stimulus-Related Artifacts

Whatever stimuli you use, you should make sure they will not cause electrical artifacts that will be picked up by the electrodes or cause reflexes that will contaminate the data. For example, most audio headphones operate by passing a current through a coil to cause the movement of a membrane with respect to a magnet, and this can induce a current in the electrodes. Thus, shielded headphones may be important for auditory ERP studies. In addition, sudden noises can elicit a muscle twitch called the post-auricular reflex, which the EEG electrodes can pick up. Somatosensory stimulation is most often produced with small shocks, and these can also induce artifacts in the electrodes and elicit reflexes.

Stimulus Timing on CRTs

The remainder of this chapter will focus on presenting visual stimuli on a computer-controlled cathode-ray tube (CRT) monitor (see Brainard, Pelli, & Robson, 2002 for a more detailed discussion). There are three reasons for focusing on visual stimuli. First, visual stimuli are used much more commonly than stimuli in other modalities. Second, the presentation of visual stimuli on a CRT involves some subtle timing issues that are particularly important in ERP experiments. Finally, I just don't have much experience with stimuli in other modalities, so I can't provide much detailed information. You can learn about the details of non-visual stimulation from the literature or by asking experts.

CRT Basics CRTs pose special problems for stimulus timing, and you need to understand how computer-controlled CRTs operate in order to avoid timing errors. As figure 8.3A illustrates, CRTs operate by passing a narrow electron beam through a magnetic field, which directs it to a specific location in a phosphor layer at the front of the monitor. When the electron beam hits the phosphor, the phosphor emits light that is proportional to the intensity of the electron beam. The location that is struck by the electron beam at a given time corresponds to a single pixel on the monitor. A color CRT uses three beams pointed at slightly different locations containing phosphors that emit red, green, and blue light; for the sake of simplicity, however, we will assume a single electron beam except as noted.

The electron beam activates only one pixel at a given moment. The magnetic field is modulated moment by moment to control which pixel is being activated. In an oscilloscope, the electron beam can be moved around in any desired pattern. A CRT, however, uses a *raster* pattern, and the electron beam is called a *raster beam* (see figure 8.3B). The raster beam starts in the upper left corner of the screen and illuminates all of the pixels on the top row of the monitor, one by one, with an intensity that is adjusted individually for each pixel. It then shifts down to the next row and draws the next set of pixels, and so on until it reaches the bottom right corner, at which point every pixel on the monitor will have been illuminated by some amount. The beam is then turned off briefly and moves back to the upper left corner of the monitor, and the whole process is repeated. The rate of repetition (the *refresh* rate) is usually between 50 and 100 Hz, corresponding to a time between refreshes of between 20 and 10 ms, respectively.

The intensity of the raster beam for each pixel is determined by a series of numbers stored in the video card's *frame buffer*. The frame buffer is just a contiguous set of memory locations, where each location contains the intensity value for a given pixel (or three values in the case of a color CRT, one each for the red, green,

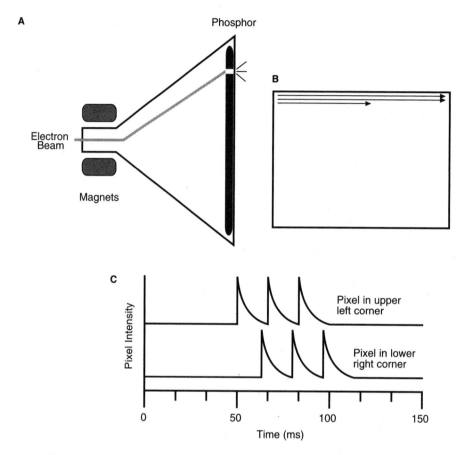

Figure 8.3   Operation of a CRT monitor. (A) The electron beam is directed by a magnetic field to a specific location at the front of the monitor, causing a small spot in the phosphor layer to become illuminated. (B) Raster pattern of the movement of the electron beam over time. (C) Intensity of the luminance from pixels on two different locations on a CRT monitor.

and blue intensities). When a program draws an image to the monitor, it is really just updating values in the frame buffer. No change occurs on the monitor until the raster beam reaches the set of pixels corresponding to the updated values, at which point the values are read from the frame buffer and the pixels are activated accordingly. This is a big part of why timing is so tricky for CRTs: the time at which something actually appears on the screen is determined by both the contents of the frame buffer, which your program directly controls, and the position of the raster beam, which your program does not directly control.

CRT Timing   As illustrated in figure 8.3C, a pixel's intensity increases almost instantaneously when the raster beam hits the phosphor, and then the intensity falls fairly rapidly toward zero when the raster beam moves away As an example, consider a scenario in which the refresh rate is set at 60 Hz (16.67 ms per refresh), the raster beam starts at the upper left corner at time zero, and the entire screen is to be turned on at 50 ms, remain visible for 50 ms, and then be turned off. The top trace in figure 8.3C shows the intensity of the pixel at the top left corner of the screen for this scenario. The intensity of the raster beam will be low when it hits this pixel at times 0, 16.67 ms, and 33.33 ms, and then transition to high when it hits the pixel at times 50, 66.67, and 83.33 ms, returning again to a low value at 100 ms. The pixel's illumination will spike upward at these times and then decay downward in the intervening periods. We don't perceive it as flickering, however, because the retina integrates information over time.[1]

Less than a microsecond after the pixel in the upper left corner is illuminated at 50 ms, the next pixel to the right will be illuminated, and then the next one, and so on until the entire top row has been illuminated (drawing one row of pixels will take approximately 120 µs with a 640 × 480 resolution). The pixels in the next row will then be sequentially illuminated, and then the next row, and so on until the entire display has been illuminated (which will take approximately 16 ms with a 60-Hz refresh rate). The bottom trace in figure 8.3C shows the actual illumination that would be observed for a pixel near the bottom of the display. The illumination of this pixel is delayed by approximately 16 ms compared to the pixel in the upper left hand corner of the display. Thus, the display is not drawn in a single instant, but instead is gradually painted over a period of many milliseconds.

Timing Errors   If your stimulus presentation program draws an object at some random time, it may cause a large, visible artifact known as *tearing*. This happens when the frame buffer is being updated in the same set of locations that are currently being drawn to the display by the raster beam. It is important to avoid this artifact, which can be very distracting.

To avoid tearing and permit precise timing, video cards send a signal to the computer called a *vertical retrace interrupt* or *video blanking interrupt* just after the raster beam has finished a cycle of drawing and just before the next cycle is about to begin. In most cases, the best way to ensure correct timing is to use the following sequence of events in the stimulus presentation program: (a) wait for the interrupt signal, (b) send an event code, and (c) draw the stimuli in a manner that ensures that any changes in the intensity values stored by the frame buffer occur after the interrupt but before those values are read by the raster beam.

By following this sequence, you can be sure that your event code is linked to the position of the electrode beam and therefore to the moment at which the illumination of the monitor actually changes. There will be a delay between the event code and the illumination of a given pixel, and this delay will depend on the location of the pixel (i.e., it will be larger for lower rows than for higher rows). However, this delay is constant for a given location, and you can determine the amount of delay by recording the illumination at a given location with the device shown in figure 8.2C. A known, constant delay is a minor annoyance, whereas an unknown, variable delay can be a major problem.

The tricky part of this sequence is the last part, making sure that you do not change values in the frame buffer after those values have already been used to control the raster beam on a given raster cycle. There are two common ways this can occur. First, if you are drawing a complex stimulus or have a slow computer or software system, it may take longer than one raster cycle to draw the stimulus. Second, you may not draw from top to bottom, and a portion of the stimulus display may be drawn to the frame buffer after the raster beam has already drawn this portion of the display. For example, imagine you are conducting a visual search experiment in which twenty-four squares are drawn at random locations across the display. If you wait for the vertical retrace interrupt and then start drawing the squares to the frame buffer in random order, you may end up drawing a square near the top of the frame buffer 5 ms after the interrupt. By this point, the raster beam will have already drawn the top portion of the frame buffer, so you may see tearing artifacts, and the square will not appear until early in the next raster cycle.

In most cases, the best way to avoid these problems is to predraw each stimulus display in an offscreen memory buffer. That is, prior to each trial of the experiment, you can predraw the stimulus displays in an area of memory that simulates the frame buffer but is not actually visible. When it is time to display the stimulus, you will wait for the vertical retrace interrupt, send an event code, and then copy the simulated frame buffer into the actual frame buffer. As long as the copying is done from top to bottom and is faster than the raster beam, this will guarantee that your timing is perfectly precise (although pixels near the bottom of the display will still be illuminated later than pixels near the top of the display). Alternatively, some video cards contain multiple frame buffers, and it is possible to predraw the stimuli in one frame buffer while another frame buffer is being displayed. The predrawn frame buffer can then be displayed by telling the video card to switch the displayed location, which is virtually instantaneous. No copying from memory into the frame buffer is necessary in this case.

Software Packages Various stimulus presentation programs are available, and some are designed expressly for use in ERP experiments. Before purchasing a program, you should inquire about the nature of the video timing. Every vendor will tell you that the timing is accurate, but you should find out exactly how the program works. In particular, you should find out whether the program (a) synchronizes stimulus presentation and event codes to the vertical retrace interrupt, (b) predraws each display to an offscreen memory buffer, and (c) might be occasionally interrupted by other programs or the operating system, which may cause large timing errors.

If you are inclined to write your own visual stimulus presentation programs, I would highly recommend using MATLAB and a set of routines called the PsychToolbox. This toolbox was developed by two highly respected vision researchers, David Brainard and Dennis Pelli (Brainard, 1997; Pelli, 1997), and they have rigorously tested its timing. MATLAB provides an excellent environment for the relatively rapid development of new experiments, and the PsychToolbox makes it easy to implement the sequence of procedures described here.

CRTs versus LCDs At the time of this writing, CRT monitors are quickly being displaced by LCD monitors, and this change may obviate much of the information provided so far. Unfortunately, LCD technology has not stabilized enough for me to provide useful information about how to achieve precise timing with LCDs.

LCD displays operate by using electrical signals to change the polarization of display elements, which in turn influences the transmission of light. No raster beam is necessary, so LCDs can potentially make instantaneous changes to pixels in any position on the screen. However, LCDs are being integrated into computer systems that were originally designed with CRTs in mind, and information from the video card's frame buffer is typically transmitted serially to the LCD display. Thus, it is still important to consider carefully the timing of the stimuli with respect to the event codes.

The optical properties of LCDs can also be problematic. In the context of ERP recordings, the biggest problem is that the transmission of light by LCDs changes relatively slowly. As a result, stimulus onsets and offsets will be more gradual than on CRTs. Fortunately, LCD manufacturers are motivated to reduce this problem because of the growing use of LCDs for television displays, video games, and other high-speed applications. Thus, LCDs may soon have better temporal properties.

To examine LCD timing firsthand, I used the circuit shown in figure 8.2 to measure the output of a high quality LCD monitor, and I recorded this signal with an ERP digitization system. Rather than using a digital LCD interface, I connected the LCD display to the computer's analog VGA output, which is designed for CRTs. Most current LCDs have both analog and digital inputs, and the digital

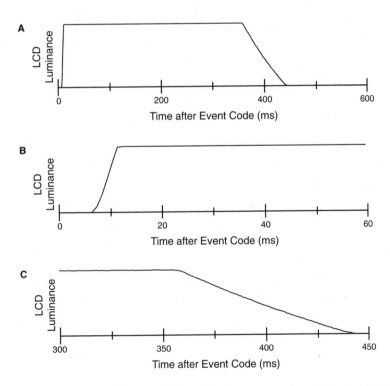

Figure 8.4 Luminance recorded from an LCD monitor in response to a white square presented on a black background. The same luminance signal is shown on three different time scales.

input will usually have higher fidelity (but may have unknown timing properties). Figure 8.4 illustrates the output of the LCD at the location of a white square that was presented on a black background. The square was supposed to onset at 0 ms (the time of the event code) and offset at 350 ms (and this is exactly what would have happened if I had used a CRT monitor).

Figure 8.4A shows the entire time course of the LCD output; the onset and offset appear on expanded time scales in figures 8.4B and 8.4C, respectively. The figure illustrates three potential problems with LCDs. First, the LCD's output is slightly delayed: The onset of the luminance change does not begin until approximately 7 ms after the event code, and the luminance does not begin to decline until approximately 357 ms. I suspect that this is caused by the analog VGA connection; the analog signal must be converted back into a digital form by the LCD monitor. Second, the onset is not instantaneous, but instead builds up over a period of approximately 5 ms. This is probably fast enough for the majority of ERP experiments, but this particular LCD display was chosen for its rapid onset time, and other LCD displays ramp up over a substantially longer period. Third, the offset of the display is very gradual, requiring almost 100 ms to reach the baseline luminance value. This won't be a significant problem when stimuli are presented on a black background. However, if the stimuli are presented on a white background, the onset of a stimulus will be achieved by a decrease in luminance, and the luminance will change in the slow manner shown in figure 8.4C. This could be quite problematic for experiments in which precise timing is important.

LCDs do have a very significant advantage over CRTs: They don't pump a stream of electrons directly toward the subject's head. Consequently, LCDs should produce less electrical noise than CRTs, and a Faraday cage may not be necessary. LCDs may therefore become the best choice for most experiments once the technology reaches maturity. In the meantime, you may wish to try using LCD monitors, but you should carefully test the timing of an each display before using it (different models may have radically different temporal properties).

Appendix: Basic Principles of Electricity

This appendix describes some important aspects of electricity and magnetism that arise when considering the neural origins of ERPs and sources of noise that can arise in ERP recordings.

Electricity is simply the flow of charges through a conductive medium. In electrical circuits, it is usually electrons that actually flow. In the nervous system, much of the electricity is due to the movement of small ions across cell membranes. But the principles of electricity are the same in both of these situations.

Voltage, Current, and Resistance

The three most fundamental terms in electricity are *voltage*, *current*, and *resistance*. Voltage is essentially electrical pressure and is analogous water pressure. Voltage is also called electrical *potential*, because it reflects the potential for electrical current to flow from one place to another. This can be understood by analogy to the flow of water through pipes. Consider, for example, a tank of water at the top of a hill. There is a lot of potential for the water to flow to the bottom of the hill, but little potential for the water to float up to the top. Importantly, the potential for water to flow downhill is present even if no water is flowing at a given moment (e.g., because a valve is closed). Similarly, there is potential for electrical current to flow from one terminal of a car battery to the other even if no current is flowing. Voltage is usually labeled *E* for *electromotive force*.

Current is the number of charged particles (e.g., electrons) that flow past a given point in a specific amount of time. Current is measured in amperes, where 1 ampere is equal to 1 coulomb

(6.24×10^{18}) of charges moving past a single point in one second. Measuring electrical current is analogous to measuring the quantity of water that passes through a given segment of pipe in a fixed time period (e.g., 10 liters per minute). Current is usually labeled I for *intensity*.

Resistance is the ability of a substance to keep charged particles from passing (it's the inverse of *conductance*). Three main factors contribute to resistance: (1) the composition of the substance, (2) its length, and (3) its diameter. Due to their molecular properties, some substances conduct electricity better than others (e.g., copper is a better conductor than zinc). However, the ability of any substance to conduct electricity will be reduced if it is very thin. To use yet another hydraulic example, consider a water filtration system in which the water supply for a house passes through a large tank filled with carbon. If the carbon is tightly packed, water will not easily flow through the tank, but if the carbon is loosely packed, water will flow easily. This is analogous to the dependence of electrical resistance on the properties of the substance. Now imagine water passing through a hose. If the hose is very long and narrow, a great deal of pressure will be necessary to fill a bucket in a short amount of time; if the hose is short and wide, the bucket will fill quickly with only a moderate amount of water pressure. This is analogous to the dependence of electrical resistance on the length and diameter of the conductor. Resistance is measured in *Ohms* (Ω) and is usually labeled R.

This last analogy also illustrates the relationships among voltage, current, and resistance. If a thin hose is used, the volume of water that passes out of the end of the hose will be small relative to what would be obtained with a wider hose and the same water pressure. Similarly, if voltage stays the same and the resistance increases, the current will decrease. However, if a thin hose is used, a large volume of water can be obtained in a given time period by increasing the water pressure. Similarly, it is possible to maintain a constant current when the resistance is increased by increasing the voltage.

Ohm's Law

These relationships are summarized by Ohm's law: $E = IR$ (voltage is equal to the product of the current and the resistance). This means that 1 volt of electromotive force is required to pass 1 ampere of current through a resistance of 1 ohm. This equation implies that if the voltage is increased, then this must be accompanied by an increase in current, an increase in resistance, or changes in both current and resistance. In particular, if the voltage increases and the resistance is constant, then the current must increase in proportion with the voltage. This makes sense, because an increase in the pressure (voltage) naturally leads to an increase in current. However, a somewhat less intuitive consequence of Ohm's law is that if the resistance increases and the current is held constant, then the voltage must increase in proportion to the resistance. This might seem counterintuitive, because you might expect that an increase in resistance would lead to a decrease in voltage, but it actually leads to an increase in voltage (assuming that current remains constant). However, this makes sense if you think about the hydraulic analogy: If you use a thinner hose but still have the same amount of water coming out the end, you must have increased the water pressure.

Another implication of Ohm's law is that it is possible to have a very large voltage without any significant current if the resistance is near infinity, which is very common (as in the case of the terminals of a car battery when they are not connected to anything). However, the only way to get a significant current without any significant voltage is to have a resistance that is near zero, which is very uncommon (it requires supercooling).

Impedance

There is one other term that I would like to mention in this context, namely *impedance*. Technically, the term *resistance* applies only when the current is constant over time (which is called direct current or DC), and *impedance* is the appropriate term to use when the current varies over time (alternating current or AC). Because ERPs vary over time, impedance is generally the most relevant concept. A different term is necessary because there are certain factors that contribute to impedance that do not contribute to DC resistance (specifically, inductance and capacitance). However, for most practical purposes, impedance is analogous to resistance, so you don't need to worry about the differences. Impedance is usually labeled Z, and most impedance meters measure the impedance using a small sine-wave voltage oscillating at around 10 Hz.

Electricity and Magnetism

Electricity and magnetism are fundamentally related to each other, and it is important to understand this relationship to understand how ERP recordings pick up electrical noise and how MEG/ERMF recordings are related to EEG/ERP recordings. Current flowing through a conductor generates a magnetic field that flows around the conductor. Moreover, if a magnetic field passes through a conductor, it induces an electrical current. Figure A.1 illustrates these

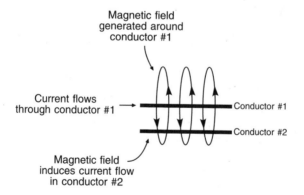

Figure A.1   Relationship between electricity and magnetism. A current is passed through conductor #1, and this generates a magnetic field that circles around conductor #1. As this magnetic field passes through conductor #2, it induces a small current in conductor #2.

two principles, showing what happens when a current passes through one of two nearby conductors. The flow of current through one of the conductors generates a magnetic field, which in turn induces current flow in the other conductor. This is how electrical noise in the environment can induce electrical activity in an ERP subject, in the electrodes, or in the wires leading from the electrodes to the amplifier.

Notes

Chapter 1

1. The concept of noise is fundamental to any data recording technique, but this term is unfamiliar to many beginners. The term noise simply refers to any source of variation in the data that is unrelated to thing you are trying to record (which is termed the signal). Keep in mind that one researcher's noise may be another researcher's signal. For example, electrical activity generated by the heart (the EKG) may sometimes contaminate ERP recordings, and ERP researchers treat it as noise. To a cardiologist, however, the EKG is a signal, not noise. Noise may be random and unpredictable or it may be nonrandom and predictable. As long as it causes the recorded data to differ from the signal, however, it's still noise.

2. This assumes that the recording electrode is at a sufficient distance from the activated cortex, which will always be true in scalp recordings.

3. To be more precise, the high resistance of the skull causes a small amount of volume-conducted electrical current to flow perpendicular to the skull, and this perpendicular electrical activity is accompanied by a small magnetic field. In this manner, the high resistance of the skull may indirectly cause a small amount of distortion in magnetic fields. However, this is a relatively minor effect compared to the distortion of the electrical activity.

4. You may wonder why all this research has not led to a consensus about the psychological or neurophysiological processes reflected by the P3 wave. I suspect there are two main contributing factors. One factor is that the P3 wave is present, at least to some extent, in almost every condition of almost every experiment. Because the P3 wave is so ubiquitous, it's hard to narrow down the general cognitive domain associated with it. A second factor is that it's difficult to isolate the P3 wave from other overlapping ERP components. It seems very likely that many different positive-going ERP components are active over the broad time range and scalp distribution of the P3 wave, each reflecting a different cognitive process. Because we don't have a good means of isolating these different P3 subcomponents, it's very hard to associate this amalgam of P3 activity with a specific psychological process.

Chapter 2

1. An abbreviated version of this chapter has been published previously (Luck, 2005).

2. This definition has a bit of wiggle room in the definition of a neuroanatomical module. You could use Brodmann's areas to define the modules, but you could also use the finer

divisions of modern neuroanatomy (Felleman & Van Essen, 1991) or coarser divisions such as the dorsal stream or the ventral stream.

3. The results in the figure are summarized in a bar graph, and ERP waveforms are not shown. Because it is difficult to measure the latent components from the observed ERP waveforms, it is important to show the ERP waveforms and not just provide measurements of component amplitudes and latencies (see the first section of chapter 6 for more details). However, at the time of this experiment, computers were quite primitive, and computing something as straightforward as the average waveform across subjects was not terribly common. The original journal article describing this experiment did not contain any grand averages, and instead showed examples of the waveforms from a few subjects. The individual-subject waveforms were just too ugly to show here.

Chapter 4

1. Technically speaking, covariance is used rather than correlation, which means that the matching EEG segment must be large as well as being similar in shape to the template.
2. The data shown in this figure were actually computed using a somewhat different technique, which used *wavelet* transforms instead of Fourier transforms. However, the general principle is the same.
3. This framework was originally developed by Jon Hansen at UCSD. He also developed several of the specific artifact rejection algorithms described later in this chapter.
4. Although a chinrest seems like it ought to reduce muscle tension, my lab has found that they actually increase muscle noise. But it may be possible to set up a chinrest in a way that reduces rather than increases muscle noise.

Chapter 5

1. In this context, power is simply amplitude squared.
2. The discussion of filtering presented here applies to transient waveforms from virtually any source, including event-related magnetic fields. However, these principles are relevant primarily for the transient responses that are elicited by discrete stimuli and not to the steady-state responses that are elicited by fast, repetitive streams of stimuli presented at a constant rate.
3. Plots of phase are often difficult to understand, and the Fourier transforms shown in this chapter will show the amplitude information without the corresponding phase information.
4. The process is actually somewhat more complex than this due to the fact that both the ERP waveform and the impulse response function are finite in duration and digitally sampled.
5. A 12.5-Hz half-amplitude cutoff is much lower than the cutoff frequencies typically employed for attenuating high frequency noise. However, filter-induced distortions are more apparent in this context with lower cutoffs, and a 12.5 Hz cutoff was therefore chosen to make these distortions clearer.
6. Artifact rejection is a nonlinear process, and filtering before artifact rejection and averaging will not yield the same result as filtering after artifact rejection and averaging

(although the result will be very similar if the specific filter used doesn't influence the artifact rejection process very much).

Chapter 6

1. Occasionally, the waveform will not have a local peak within the measurement time range that exceeds the average of the three to five points on either side. This occurs when the waveform gradually increases or decreases monotonically through the entire measurement window. In this case, you can just use the simple peak amplitude or the amplitude at the midpoint of the latency range. These aren't perfect solutions, but they're better than always using the simple peak amplitude.

2. The Greenhouse-Geisser adjustment tends to be overly conservative, especially at moderate to high levels of non-sphericity. Many statistical packages also include the Hyun-Feldt adjustment, which is probably somewhat more reasonable and may be useful when low statistical power is a problem.

Chapter 7

1. A physicist with no knowledge of BESA or ERPs was also tested, but his localizations were both unrepresentative and far from correct, so I will not include his results in the discussion here.

2. If two dipoles are assumed to be mirror-symmetrical, only seven parameters are required to represent the two dipoles, rather than the twelve that would be otherwise required. That is, six parameters are used to represent one of the two dipoles, as usual, but the other dipole requires only one additional parameter, representing its magnitude (which is the only parameter that may differ from the other dipole). Moreover, because magnitude is treated differently from the other parameters in BESA, the assumption of mirror symmetry essentially removes all of the major parameters from one of the two dipoles. Thus, this ten-dipole simulation is equivalent to an unconstrained seven-dipole simulation.

Chapter 8

1. Because of the integration time of the retina, reducing the duration of a stimulus from 100 ms to some shorter duration leads to the perception of a dimmer stimulus rather than a shorter stimulus (this is *Bloch's law*). In addition, stimuli of approximately 100 ms or less do not elicit separate onset and offset ERP responses, but longer stimuli do. Consequently, I would recommend a duration of 100 ms for experiments with brief stimuli, and I would recommend a duration that exceeds the interesting portion of the ERP waveform for experiments with long stimuli. This way you will not have offset responses in the middle of your waveform.

References

Adrian, E. D., & Matthews, B. H. C. (1934). The Berger rhythm: Potential changes from the occipital lobes in man. *Brain, 57*, 355–385.

Aine, C., Huang, M., Stephen, J., & Christner, R. (2000). Multistart algorithms for MEG empirical data analysis reliably characterize locations and time courses of multiple sources. *Neuroimage, 12*, 159–172.

Alcaini, M., Giard, M. H., Thevenet, M., & Pernier, J. (1994). Two separate frontal components in the N1 wave of the human auditory evoked response. *Psychophysiology, 31*, 611–615.

American Encephalographic Society. (1994). Guidelines for standard electrode position nomenclature. *Journal of Clinical Neurophysiology, 11*, 111–113.

Baillet, S., Mosher, J. C., Leahy, R. M., & Shattuck, D. W. (1999). BrainStorm: A matlab toolbox for the processing of MEG and EEG signals. *Neuroimage, 9*, 246.

Bentin, S., Allison, T., Puce, A., Perez, E., & McCarthy, G. (1996). Electrophysiological studies of face perception in humans. *Journal of Cognitive Neuroscience, 8*, 551–565.

Bentin, S., & Carmel, D. (2002). Accounts for the N170 face-effect: A reply to Rossion, Curran, & Gauthier. *Cognition, 85*, 197–202.

Berg, P., & Scherg, M. (1991a). Dipole modelling of eye activity and its application to the removal of eye artefacts from the EEG and MEG. *Clinical Physics & Physiological Measurement, 12 Suppl A*, 49–54.

Berg, P., & Scherg, M. (1991b). Dipole models of eye movements and blinks. *Electroencephalography & Clinical Neurophysiology, 79*, 36–44.

Berg, P., & Scherg, M. (1994). A multiple source approach to the correction of eye artifacts. *Electroencephalography & Clinical Neurophysiology, 90*, 229–241.

Berger, H. (1929). Ueber das Elektrenkephalogramm des Menschen. *Archives fur Psychiatrie Nervenkrankheiten, 87*, 527–570.

Bertrand, O., Perrin, F., & Pernier, J. (1991). Evidence for a tonotopic organization of the auditory cortex with auditory evoked potentials. *Acta Otolaryngologica, 491*, 116–123.

Brainard, D. H. (1997). The psychophysics toolbox. *Spatial Vision, 10*, 433–436.

Brainard, D. H., Pelli, D. G., & Robson, T. (2002). Display characterization. In J. Hornak (Ed.), *Encyclopedia of Imagine Science and Technology* (pp. 172–188). New York: Wiley.

Brandeis, D., Naylor, H., Halliday, R., Callaway, E., & Yano, L. (1992). Scopolamine effects on visual information processing, attention, and event-related potential map latencies. *Psychophysiology, 29*, 315–336.

Broadbent, D. E. (1958). *Perception and Communication*. New York: Pergamon.

Carmel, D., & Bentin, S. (2002). Domain specificity versus expertise: Factors influencing distinct processing of faces. *Cognition, 83*, 1–29.

Chelazzi, L., Duncan, J., Miller, E. K., & Desimone, R. (1998). Responses of neurons in inferior temporal cortex during memory-guided visual search. *Journal of Neurophysiology, 80*, 2918–2940.

Chelazzi, L., Miller, E. K., Duncan, J., & Desimone, R. (1993). A neural basis for visual search in inferior temporal cortex. *Nature, 363*, 345–347.

Chelazzi, L., Miller, E. K., Duncan, J., & Desimone, R. (2001). Responses of neurons in macaque area V4 during memory-guided visual search. *Cerebral Cortex, 11*, 761–772.

Chun, M. M., & Potter, M. C. (1995). A two-stage model for multiple target detection in rapid serial visual presentation. *Journal of Experimental Psychology: Human Perception and Performance, 21*, 109–127.

Clark, V. P., Fan, S., & Hillyard, S. A. (1995). Identification of early visually evoked potential generators by retinotopic and topographic analyses. *Human Brain Mapping, 2*, 170–187.

Cohen, D., & Cuffin, B. N. (1991). EEG versus MEG localizaiton accuracy: Theory and experiment. *Brain Topography, 4*, 95–103.

Coles, M. G. H. (1989). Modern mind-brain reading: Psychophysiology, physiology and cognition. *Psychophysiology, 26*, 251–269.

Coles, M. G. H., Gratton, G., Kramer, A. F., & Miller, G. A. (1986). Principles of signal acquisition and analysis. In M. G. H. Coles, E. Donchin & S. W. Porges (Eds.), *Psychophysiology: Systems, Processes, and Applications* (pp. 183–221). New York: Guilford Press.

Coles, M. G. H., & Rugg, M. D. (1995). Event-related potentials: An introduction. In M. D. Rugg & M. G. H. Coles (Eds.), *Electrophysiology of Mind* (pp. 1–26). New York: Oxford University Press.

Coles, M. G. H., Smid, H., Scheffers, M. K., & Otten, L. J. (1995). Mental chronometry and the study of human information processing. In M. D. Rugg & M. G. H. Coles (Eds.), *Electrophysiology of Mind: Event-Related Brain Potentials and Cognition* (pp. 86–131). Oxford: Oxford University Press.

Courchesne, E., Hillyard, S. A., & Galambos, R. (1975). Stimulus novelty, task relevance and the visual evoked potential in man. *Electroencephalography and Clinical Neurophysiology, 39*, 131–142.

Cuffin, B. N., Cohen, D., Yunokuchi, K., Maniewski, R., Purcell, C., Cosgrove, G. R., Ives, J., Kennedy, J., & Schomer, D. (1991). Test of EEG localization accuracy using implanted sources in the human brain. *Annals of Neurology, 29*, 132–138.

Dale, A. M., & Sereno, M. I. (1993). Improved localization of cortical activity by combining EEG and MEG with MRI cortical surface reconstruction: A linear approach. *Journal of Cognitive Neuroscience, 5*, 162–176.

Davis, H., Davis, P. A., Loomis, A. L., Harvey, E. N., & Hobart, G. (1939). Electrical reactions of the human brain to auditory stimulation during sleep. *Journal of Neurophysiology, 2*, 500–514.

Davis, P. A. (1939). Effects of acoustic stimuli on the waking human brain. *Journal of Neurophysiology, 2*, 494–499.

de Jong, R., Coles, M. G., Logan, G. D., & Gratton, G. (1990). In search of the point of no return: The control of response processes. *Journal of Experimental Psychology: Human Perception & Performance, 16*, 164–182.

Dehaene, S., Naccache, L., Le Clec, H. G., Koechlin, E., Mueller, M., Dehaene-Lambertz, G., van de Moortele, P. F., & Le Bihan, D. (1998). Imaging unconscious semantic priming. *Nature, 395*, 597–600.

Dehaene, S., Posner, M. I., & Tucker, D. M. (1994). Localization of a neural system for error detection and compensation. *Psychological Science, 5*, 303–305.

Delorme, A., & Makeig, S. (2004). EEGLAB: An open source toolbox for analysis of single-trial EEG dynamics including independent component analysis. *Journal of Neuroscience Methods, 134*, 9–21.

Desmedt, J. E., Chalklin, V., & Tomberg, C. (1990). Emulation of somatosensory evoked potential (SEP) components with the 3-shell head model and the problem of "ghost potential fields" when using an average reference in brain mapping. *Electroencephalography & Clinical Neurophysiology, 77*, 243–258.

Deutsch, J. A., & Deutsch, D. (1963). Attention: Some theoretical considerations. *Psychological Review, 70*, 80–90.

Di Russo, F., Martinez, A., Sereno, M. I., Pitzalis, S., & Hillyard, S. A. (2002). Cortical sources of the early components of the visual evoked potential. *Human Brain Mapping, 15*, 95–111.

Di Russo, F., Teder-Sälejärvi, W. A., & Hillyard, S. A. (2003). Steady-state VEP and attentional visual processing. In A. Zani & A. M. Proverbio (Eds.), *The Cognitive Electrophysiology of Mind and Brain* (pp. 259–274). San Diego: Academic Press.

Dien, J. (1998). Issues in the application of the average reference: Review, critiques, and recommendations. *Behavior Research Methods, Instruments, & Computers, 30*, 34–43.

Donchin, E. (1981). Surprise! . . . Surprise? *Psychophysiology, 18*, 493–513.

Donchin, E. (1979). Event-related brain potentials: A tool in the study of human information processing. In H. Begleiter (Ed.), *Evoked Brain Potentials and Behavior* (Vol. 2, pp. 13–88). New York: Plenum Press.

Donchin, E., & Coles, M. G. H. (1988). Is the P300 component a manifestation of context updating? *Behavioral Brain Science, 11*, 357–374.

Donchin, E., & Heffley, E. F., III. (1978). Multivariate analysis of event-related potential data: A tutorial review. In D. Otto (Ed.), *Multidisciplinary Perspectives in Event-Related Brain Potential Research* (pp. 555–572). Washington, D.C.: U.S. Government Printing Office.

Duncan-Johnson, C., & Donchin, E. (1979). The time constant in P300 recording. *Psychophysiology, 16*, 53–55.

Duncan-Johnson, C. C., & Donchin, E. (1977). On quantifying surprise: The variation of event-related potentials with subjective probability. *Psychophysiology, 14*, 456–467.

Duncan-Johnson, C. C., & Kopell, B. S. (1981). The Stroop effect: Brain potentials localize the source of interference. *Science, 214*, 938–940.

Eimer, M. (1996). The N2pc component as an indicator of attentional selectivity. *Electroencephalography and Clinical Neurophysiology, 99*, 225–234.

Eriksen, C. W., & Schultz, D. W. (1979). Information processing in visual search: A continuous flow conception and experimental results. *Perception and Psychophysics, 25*, 249–263.

Falkenstein, M., Hohnsbein, J., Joormann, J., & Blanke, L. (1990). Effects of errors in choice reaction tasks on the ERP under focused and divided attention. In C. H. M. Brunia, A. W. K. Gaillard & A. Kok (Eds.), *Psychophysiological Brain Research* (pp. 192–195). Amsterdam: Elsevier.

Felleman, D. J., & Van Essen, D. C. (1991). Distributed hierarchical processing in the primate cerebral cortex. *Cerebral Cortex, 1*, 1–47.

Ferree, T. C., Luu, P., Russell, G. S., & Tucker, D. M. (2001). Scalp electrode impedance, infection risk, and EEG data quality. *Clinical Neurophysiology, 112*, 536–544.

Gaillard, A. W. K. (1988). Problems and paradigms in ERP research. *Biological Psychology, 26*, 91–109.

Galambos, R., & Sheatz, G. C. (1962). An electroencephalographic study of classical conditioning. *American Journal of Physiology, 203*, 173–184.

Ganis, G., Kutas, M., & Sereno, M. I. (1996). The search for "common sense": An electrophysiological study of the comprehension of words and pictures in reading. *Journal of Cognitive Neuroscience, 8*, 89–106.

Garner, W. R., Hake, H. W., & Eriksen, C. W. (1956). Operationism and the concept of perception. *Psychology Review, 63*, 149–159.

Gehring, W. J., Goss, B., Coles, M. G. H., Meyer, D. E., & Donchin, E. (1993). A neural system for error-detection and compensation. *Psychological Science, 4*, 385–390.

Gehring, W. J., & Willoughby, A. R. (2002). The medial frontal cortex and the rapid processing of monetary gains and losses. *Science, 295*, 2279–2282.

George, J. S., Aine, C. J., Mosher, J. C., Schmidt, D. M., Ranken, D. M., Schlitt, H. A., Wood, C. C., Lewine, J. D., Sanders, J. A., & Belliveau, J. W. (1995). Mapping function in the human brain with magnetoencephalography, anatomical magnetic resonance imaging, and functional magnetic resonance imaging. *Journal of Clinical Neurophysiology, 12*, 406–431.

George, N., Evans, J., Fiori, N., Davidoff, J., & Renault, B. (1996). Brain events related to normal and moderately scrambled faces. *Cognitive Brain Research, 4*, 65–76.

Gevins, A., Le, J., Leong, H., McEvoy, L. K., & Smith, M. E. (1999). Deblurring. *Journal of Clinical Neurophysiology, 16*, 204–213.

Gibbs, F. A., Davis, H., & Lennox, W. G. (1935). The electro-encephalogram in epilepsy and in conditions of impaired consciousness. *Archives of Neurology and Psychiatry, 34*, 1133–1148.

Girelli, M., & Luck, S. J. (1997). Are the same attentional mechanisms used to detect visual search targets defined by color, orientation, and motion? *Journal of Cognitive Neuroscience, 9*, 238–253.

Glaser, E. M., & Ruchkin, D. S. (1976). *Principles of Neurobiological Signal Analysis.* New York: Academic Press.

Gratton, G., Coles, M. G. H., & Donchin, E. (1983). A new method for off-line removal of ocular artifacts. *Electroencephalography and Clinical Neurophysiology, 55*, 468–484.

Gratton, G., Coles, M. G. H., Sirevaag, E. J., Eriksen, C. W., & Donchin, E. (1988). Pre- and post-stimulus activation of response channels: A psychophysiological analysis. *Journal of Experimental Psychology: Human Perception and Performance, 14*, 331–344.

Gray, C. M., König, P., Engel, A. K., & Singer, W. (1989). Oscillatory responses in cat visual cortex exhibit inter-columnar synchronization which reflects global stimulus properties. *Nature, 338*, 334–337.

Hagoort, P., Brown, C. M., & Swaab, T. Y. (1996). Lexical-semantic event-related potential effects in patients with left hemisphere lesions and aphasia, and patients with right hemisphere leions without aphasia. *Brain, 119*, 627–649.

Hämäläinen, M. S., Hari, R., Ilmonieni, R. J., Knuutila, J., & Lounasmaa, O. V. (1993). Magnetoencephalography—theory, instrumentation, and applications to noninvasive studies of the working human brain. *Review of Modern Physics, 65*, 413–497.

Hämäläinen, M. S., & Ilmonieni, R. J. (1984). Interpreting measured magnetic frields of the brain: Estimates of current distributions. In *Technical Report TKK-F-A599*. Espoo: Helsinki University of Techhnology.

Handy, T. C., Solotani, M., & Mangun, G. R. (2001). Perceptual load and visuocortical processing: Event-related potentials reveal sensory-level selection. *Psychological Science, 12*, 213–218.

Hansen, J. C., & Hillyard, S. A. (1980). Endogenous brain potentials associated with selective auditory attention. *Electroencephalography and Clinical Neurophysiology, 49*, 277–290.

Heinze, H. J., Mangun, G. R., Burchert, W., Hinrichs, H., Scholz, M., Münte, T. F., Gös, A., Scherg, M., Johannes, S., Hundeshagen, H., Gazzaniga, M. S., & Hillyard, S. A. (1994). Combined spatial and temporal imaging of brain activity during visual selective attention in humans. *Nature, 372*, 543–546.

Helmholtz, H. (1853). Ueber einige Gesetze der Vertheilung elektrischer Ströme in körperlichen Leitern mit Anwendung auf die thierisch-elektrischen Versuche. *Annalen der Physik und Chemie, 89*, 211–233, 354–377.

Hillyard, S. A., & Galambos, R. (1970). Eye movement artifact in the CNV. *Electroencephalography and Clinical Neurophysiology, 28*, 173–182.

Hillyard, S. A., Hink, R. F., Schwent, V. L., & Picton, T. W. (1973). Electrical signs of selective attention in the human brain. *Science, 182*, 177–179.

Hillyard, S. A., & Picton, T. W. (1987). Electrophysiology of cognition. In F. Plum (Ed.), *Handbook of Physiology: Section 1. The Nervous System: Volume 5. Higher Functions of the Brain, Part 2* (pp. 519–584). Bethesda, MD: Waverly Press.

Hillyard, S. A., Vogel, E. K., & Luck, S. J. (1998). Sensory gain control (amplification) as a mechanism of selective attention: Electrophysiological and neuroimaging evidence. *Philosophical Transactions of the Royal Society: Biological Sciences, 353*, 1257–1270.

Holcomb, P. J., & McPherson, W. B. (1994). Event-related brain potentials reflect semantic priming in an object decision task. *Brain and Cognition, 24*, 259–276.

Holroyd, C. B., Niewenhuis, S., Yeung, N., Nystrom, L., Mars, R. B., Coles, M. G. H., & Cohen, J. D. (2004). Dorsal anterior cingulate cortex shows fMRI response to internal and external error signals. *Nature Neuroscience, 7*, 497–498.

Hopf, J.-M., Luck, S. J., Girelli, M., Hagner, T., Mangun, G. R., Scheich, H., & Heinze, H.-J. (2000). Neural sources of focused attention in visual search. *Cerebral Cortex, 10*, 1233–1241.

Hopf, J.-M., Vogel, E. K., Woodman, G. F., Heinze, H.-J., & Luck, S. J. (2002). Localizing visual discrimination processes in time and space. *Journal of Neurophysiology, 88*, 2088–2095.

Hopfinger, J. B., Luck, S. J., & Hillyard, S. A. (2004). Selective attention: Electrophysiological and neuromagnetic studies. In M. S. Gazzaniga (Ed.), *The Cognitive Neurosciences, Volume 3* (pp. 561–574). Cambridge, MA: MIT Press.

Hopfinger, J. B., & Mangun, G. R. (1998). Reflective attention modulates processing of visual stimuli in human extrastriate cortex. *Psychological Science, 9*, 441–447.

Huang, M., Aine, C. J., Supek, S., Best, E., Ranken, D., & Flynn, E. R. (1998). Multi-start downhill simplex method for spatio-temporal source localization in magnetoencephalography. *Electroencephalography & Clinical Neurophysiology, 108*, 32–44.

Ikui, A. (2002). A review of objective measures of gustatory function. *Acta Oto-Laryngologica Supplement, 546*, 60–68.

Ilmoniemi, R. J. (1995). Estimating brain source distributions: Comments on LORETA. *ISBET Newsletter*, *6*, 12–14.

Isreal, J. B., Chesney, G. L., Wickens, C. D., & Donchin, E. (1980). P300 and tracking difficulty: Evidence for multiple resources in dual-task performance. *Psychophysiology*, *17*, 259–273.

Ito, S., Stuphorn, V., Brown, J. W., & Schall, J. D. (2003). Performance monitoring by the anterior cingulate cortex during saccade countermanding. *Science*, *302*, 120–122.

Jasper, H. H. (1958). The ten-twenty electrode system of the International Federation. *Electroencephalography & Clinical Neurophysiology*, *10*, 371–375.

Jasper, H. H., & Carmichael, L. (1935). Elecrical potentials from the intact human brain. *Science*, *81*, 51–53.

Jeffreys, D. A. (1989). A face-responsive potential recorded from the human scalp. *Experimental Brain Research*, *78*, 193–202.

Jeffreys, D. A., & Axford, J. G. (1972). Source locations of pattern-specific components of human visual evoked potentials. I: Components of striate cortical origin. *Experimental Brain Research*, *16*, 1–21.

Jennings, J. R., & Wood, C. C. (1976). The e-adjustment procedure for repeated-measures analyses of variance. *Psychophysiology*, *13*, 277–278.

Johnson, R., Jr. (1984). P300: A model of the variables controlling its amplitude. *Annals of the New York Academy of Sciences*, *425*, 223–229.

Johnson, R., Jr. (1986). A triarchic model of P300 amplitude. *Psychophysiology*, *23*, 367–384.

Joyce, C. A., Gorodnitsky, I. F., King, J. W., & Kutas, M. (2002). Tracking eye fixations with electroocular and electroencephalographic recordings. *Psychophysiology*, *39*, 607–618.

Joyce, C. A., Gorodnitsky, I. F., & Kutas, M. (2004). Automatic removal of eye movement and blink artifacts from EEG data using blind component separation. *Psychophysiology*, *41*, 313–325.

Jung, T. P., Makeig, S., Humphries, C., Lee, T. W., McKeown, M. J., Iragui, V., & Sejnowski, T. J. (2000). Removing electroencephalographic artifacts by blind source separation. *Psychophysiology*, *37*, 163–178.

Jung, T. P., Makeig, S., Westerfield, M., Townsend, J., Courchesne, E., & Sejnowski, T. J. (2000). Removal of eye activity artifacts from visual event-related potentials in normal and clinical subjects. *Clinical Neurophysiology*, *111*, 1745–1758.

Keppel, G. (1982). *Design and Analysis*. Englewood Cliffs, NJ: Prentice-Hall.

Klem, G. H., Luders, H. O., Jasper, H. H., & Elger, C. (1999). The ten-twenty electrode system of the International Federation. *Electroencephalography & Clinical Neurophysiology, Supplement 52*, 3–6.

Koles, Z. J. (1998). Trends in EEG source localization. *Electroencephalography & Clinical Neurophysiology*, *106*, 127–137.

Kornhuber, H. H., & Deecke, L. (1965). Hirnpotentialanderungen bei Wilkurbewegungen und passiven Bewegungen des Menschen: Bereitschaftspotential und reafferente potentials. *Pflugers Archives*, *284*, 1–17.

Kramer, A. F. (1985). The interpretation of the component structure of event-related brain potentials: An analysis of expert judgments. *Psychophysiology*, *22*, 334–344.

Kutas, M. (1997). Views on how the electrical activity that the brain generates reflects the functions of different language structures. *Psychophysiology*, *34*, 383–398.

Kutas, M., & Hillyard, S. A. (1980). Reading senseless sentences: Brain potentials reflect semantic incongruity. *Science, 207*, 203–205.

Kutas, M., Hillyard, S. A., & Gazzaniga, M. S. (1988). Processing of semantic anomaly by right and left hemispheres of commissurotomy patients. *Brain, 111*, 553–576.

Kutas, M., McCarthy, G., & Donchin, E. (1977). Augmenting mental chronometry: The P300 as a measure of stimulus evaluation time. *Science, 197*, 792–795.

Leahy, R. M., Mosher, J. C., Spencer, M. E., Huang, M. X., & Lewine, J. D. (1998). A study of dipole localizaiton accuracy for MEG and EEG suing a human skull phantom. *Electroencephalography & Clinical Neurophysiology, 107*, 159–173.

Leuthold, H., & Sommer, W. (1998). Postperceptual effects and P300 latency. *Psychophysiology, 35*, 34–46.

Lindsley, D. B. (1969). Average evoked potentials—achievements, failures and prospects. In E. Donchin & D. B. Lindsley (Eds.), *Average Evoked Potentials: Methods, Results and Evaluations* (pp. 1–43). Washington, D.C.: U.S. Government Printing Office.

Lins, O. G., Picton, T. W., Berg, P., & Scherg, M. (1993a). Ocular artifacts in EEG and event-related potentials I: Scalp topography. *Brain Topography, 6*, 51–63.

Lins, O. G., Picton, T. W., Berg, P., & Scherg, M. (1993b). Ocular artifacts in recording EEGs and event-related potentials. II: Source dipoles and source components. *Brain Topography, 6*, 65–78.

Liu, A. K., Belliveau, J. W., & Dale, A. M. (1998). Spatiotemporal imaging of human brain activity using functional MRI constrained magnetoencephalography data: Monte Carlo simulations. *Proceedings of the National Academy of Sciences of the United States of America, 95*, 8945–8950.

Loveless, N. E., & Sanford, A. J. (1975). The impact of warning signal intensity on reaction time and components of the contingent negative variation. *Biological Psychology, 2*, 217–226.

Luck, S. J. (1998a). Neurophysiology of selective attention. In H. Pashler (Ed.), *Attention* (pp. 257–295). East Sussex: Psychology Press.

Luck, S. J. (1998b). Sources of dual-task interference: Evidence from human electrophysiology. *Psychological Science, 9*, 223–227.

Luck, S. J. (1999). Direct and indirect integration of event-related potentials, functional magnetic resonance images, and single-unit recordings. *Human Brain Mapping, 8*, 15–120.

Luck, S. J. (2005). Ten simple rules for designing ERP experiments. In T. C. Handy (Ed.), *Event-Related Potentials: A Methods Handbook* (pp. 17–32). Cambridge, MA: MIT Press.

Luck, S. J., Fan, S., & Hillyard, S. A. (1993). Attention-related modulation of sensory-evoked brain activity in a visual search task. *Journal of Cognitive Neuroscience, 5*, 188–195.

Luck, S. J., & Girelli, M. (1998). Electrophysiological approaches to the study of selective attention in the human brain. In R. Parasuraman (Ed.), *The Attentive Brain* (pp. 71–94). Cambridge, MA: MIT Press.

Luck, S. J., Girelli, M., McDermott, M. T., & Ford, M. A. (1997). Bridging the gap between monkey neurophysiology and human perception: An ambiguity resolution theory of visual selective attention. *Cognitive Psychology, 33*, 64–87.

Luck, S. J., & Hillyard, S. A. (1990). Electrophysiological evidence for parallel and serial processing during visual search. *Perception & Psychophysics, 48*, 603–617.

Luck, S. J., & Hillyard, S. A. (1994a). Electrophysiological correlates of feature analysis during visual search. *Psychophysiology, 31*, 291–308.

Luck, S. J., & Hillyard, S. A. (1994b). Spatial filtering during visual search: Evidence from human electrophysiology. *Journal of Experimental Psychology: Human Perception and Performance, 20,* 1000–1014.

Luck, S. J., & Hillyard, S. A. (1995). The role of attention in feature detection and conjunction discrimination: An electrophysiological analysis. *International Journal of Neuroscience, 80,* 281–297.

Luck, S. J., Hillyard, S. A., Mouloua, M., Woldorff, M. G., Clark, V. P., & Hawkins, H. L. (1994). Effects of spatial cuing on luminance detectability: Psychophysical and electrophysiological evidence for early selection. *Journal of Experimental Psychology: Human Perception and Performance, 20,* 887–904.

Luck, S. J., Vogel, E. K., & Shapiro, K. L. (1996). Word meanings can be accessed but not reported during the attentional blink. *Nature, 382,* 616–618.

Luck, S. J., Woodman, G. F., & Vogel, E. K. (2000). Event-related potential studies of attention. *Trends in Cognitive Sciences, 4,* 432–440.

Macmillan, N. A. (1999). Editorial. *Perception & Psychophysics, 61,* 1–2.

Magliero, A., Bashore, T. R., Coles, M. G. H., & Donchin, E. (1984). On the dependence of P300 latency on stimulus evaluation processes. *Psychophysiology, 21,* 171–186.

Makeig, S. (1993). Auditory event-related dynamics of the EEG spectrum and effects of exposure to tones. *Electroencephalography and Clinical Neurophysiology, 86,* 283–293.

Makeig, S., Jung, T. P., Bell, A. J., Ghahremani, D., & Sejnowski, T. J. (1997). Blind separation of auditory event-related brain responses into independent components. *Proceedings of the National Academy of Science, 94,* 10979–10984.

Mangun, G. R. (1995). Neural mechanisms of visual selective attention. *Psychophysiology, 32,* 4–18.

McCarthy, G., & Donchin, E. (1981). A metric for thought: A comparison of P300 latency and reaction time. *Science, 211,* 77–80.

McCarthy, G., Nobre, A. C., Bentin, S., & Spencer, D. D. (1995). Language-related field potentials in the anterior-medial temporal lobe: I. Intracranial distribution and neural generators. *Journal of Neuroscience, 15,* 1080–1089.

McCarthy, G., & Wood, C. C. (1985). Scalp distributions of event-related potentials: An ambiguity associated with analysis of variance models. *Electroencephalography and Clinical Neurophysiology, 62,* 203–208.

McClelland, J. L. (1979). On the time relations of mental processes: An examination of systems of processes in cascade. *Psychological Review, 86,* 287–330.

Miller, J., & Hackley, S. A. (1992). Electrophysiological evidence for temporal overlap among contingent mental processes. *Journal of Experimental Psychology: General, 121,* 195–209.

Miller, J., Patterson, T., & Ulrich, R. (1998). Jackknife-based method for measuring LRP onset latency differences. *Psychophysiology, 35,* 99–115.

Miller, J., Riehle, A., & Requin, J. (1992). Effects of preliminary perceptual output on neuronal activity of the primary motor cortex. *Journal of Experimental Psychology: Human Perception and Performance, 18,* 1121–1138.

Miltner, W., Braun, C., Johnson, R., Jr., Simpson, G. V., & Ruchkin, D. S. (1994). A test of brain electrical source analysis (BESA): A simulation study. *Electroencephalography & Clinical Neurophysiology, 91,* 295–310.

Moran, J., & Desimone, R. (1985). Selective attention gates visual processing in the extrastriate cortex. *Science, 229,* 782–784.

Mosher, J. C., Baillet, S., & Leahy, R. M. (1999). EEG source localization and imaging using multiple signal classification approaches. *Journal of Clinical Neurophysiology*, *16*, 225–238.

Mosher, J. C., & Leahy, R. M. (1999). Source localization using recursively applied and projected (RAP) MUSIC. *IEEE Transactions on Signal Processing*, *47*, 332–340.

Mosher, J. C., Lewis, P. S., & Leahy, R. M. (1992). Multiple dipole modeling and localization from spatio-temporal MEG data. *IEEE Transactions on Biomedical Engineering*, *39*, 541–557.

Näätänen, R., & Picton, T. (1987). The N1 wave of the human electric and magnetic response to sound: A review and an analysis of the component structure. *Psychophysiology*, *24*, 375–425.

Näätänen, R., & Picton, T. W. (1986). N2 and automatic versus controlled processes. In W. C. McCallum, R. Zappoli & F. Denoth (Eds.), *Cerebral Psychophysiology: Studies in Event-Related Potentials* (pp. 169–186). Amsterdam: Elsevier.

Nagamine, T., Toro, C., Balish, M., Deuschl, G., Wang, B., Sato, S., Shibasaki, H., & Hallett, M. (1994). Cortical magnetic and electrical fields associated with voluntary finger movements. *Brain Topography*, *6*, 175–183.

Norman, D. A. (1968). Toward a theory of memory and attention. *Psychological Review*, *75*, 522–536.

Nunez, P. L. (1981). *Electric Fields of the Brain*. New York: Oxford University Press.

Osman, A., Bashore, T. R., Coles, M., Donchin, E., & Meyer, D. (1992). On the transmission of partial information: Inferences from movement-related brain potentials. *Journal of Experimental Psychology: Human Perception and Performance*, *18*, 217–232.

Osman, A., & Moore, C. M. (1993). The locus of dual-task interference: Psychological refractory effects on movement-related brain potentials. *Journal of Experimental Psychology: Human Perception and Performance*, *19*, 1292–1312.

Osterhout, L., & Holcomb, P. J. (1992). Event-related brain potentials elicited by syntactic anomaly. *Journal of Memory & Language*, *31*, 785–806.

Osterhout, L., & Holcomb, P. J. (1995). Event-related potentials and language comprehension. In M. D. Rugg & M. G. H. Coles (Eds.), *Electrophysiology of Mind* (pp. 171–215). New York: Oxford University Press.

Paller, K. A. (1990). Recall and stem-completion priming have different electrophysiological correlates and are modified differentially by directed forgetting. *Journal of Experimental Psychology: Learning, Memory and Cognition*, *16*, 1021–1032.

Pascual-Marqui, R. D. (2002). Standardized low-resolution brain electromagnetic tomography (sLORETA): Technical details. *Methods & Findings in Experimental & Clinical Pharmacology*, *24 Suppl D*, 5–12.

Pascual-Marqui, R. D., Esslen, M., Kochi, K., & Lehmann, D. (2002). Functional imaging with low-resolution brain electromagnetic tomography (LORETA): A review. *Methods & Findings in Experimental & Clinical Pharmacology*, *24 Suppl C*, 91–95.

Pascual-Marqui, R. D., Michel, C. M., & Lehmann, D. (1994). Low resolution electromagnetic tomography: A new method for localizing electrical activity in the brain. *International Journal of Psychophysiology*, *18*, 49–65.

Pelli, D. G. (1997). The VideoToolbox software for visual psychophysics: Transforming numbers into movies. *Spatial Vision*, *10*, 437–442.

Pernier, J., Perrin, F., & Bertrand, O. (1988). Scalp current density fields: Concept and properties. *Electroencephalography and Clinical Neurophysiology, 69,* 385–389.

Perrin, F., Pernier, J., Bertrand, O., & Echallier, J. F. (1989). Spherical splines for scalp potential and current density mapping. *Electroencephalography and Clinical Neurophysiology, 72,* 184–187.

Phillips, C., Rugg, M. D., & Friston, K. J. (2002). Anatomically informed basis functions for EEG source localization: Combining functional and anatomical constraints. *Neuroimage, 16,* 678–695.

Picton, T. W. (1992). The P300 wave of the human event-related potential. *Journal of Clinical Neurophysiology, 9,* 456–479.

Picton, T. W., Bentin, S., Berg, P., Donchin, E., Hillyard, S. A., Johnson, R., Jr., Miller, G. A., Ritter, W., Ruchkin, D. S., Rugg, M. P., & Taylor, M. J. (2000). Guidelines for using human event-related potentials to study cognition: Recording standards and publication criteria. *Psychophysiology, 37,* 127–152.

Picton, T. W., & Hillyard, S. A. (1972). Cephalic skin potentials in electroencephalography. *Electroencephalogray and Clinical Neurophysiology, 33,* 419–424.

Picton, T. W., Hillyard, S. A., & Galambos, R. (1974). Cortical evoked responses to omitted stimuli. In M. N. Livanov (Ed.), *Basic Problems in Brain Electrophysiology* (pp. 302–311). Moscow: Nauka.

Picton, T. W., Linden, R. D., Hamel, G., & Maru, J. T. (1983). Aspects of averaging. *Seminars in Hearing, 4,* 327–341.

Picton, T. W., Lins, O. G., & Scherg, M. (1995). The recording and analysis of event-related potentials. In F. Boller & J. Grafman (Eds.), *Handbook of Neuropsychology, Vol. 10* (pp. 3–73). New York: Elsevier.

Picton, T. W., & Stuss, D. T. (1980). The component structure of the human event-related potentials. In H. H. Kornhuber & L. Deecke (Eds.), *Motivation, Motor and Sensory Processes of the Brain* (pp. 17–49). North-Holland: Elsevier.

Platt, J. R. (1964). Strong inference. *Science, 146,* 347–353.

Plonsey, R. (1963). Reciprocity applied to volume conductors and the EEG. *IEEE Transactions on Biomedical Engineering, 19,* 9–12.

Polich, J. (2004). Clinical application of the P300 event-related brain potential. *Physical Medicine & Rehabilitation Clinics of North America, 15,* 133–161.

Polich, J., & Comerchero, M. D. (2003). P3a from visual stimuli: Typicality, task, and topography. *Brain Topography, 15,* 141–152.

Polich, J., & Kok, A. (1995). Cognitive and biological determinants of P300: An integrative review. *Biological Psychology, 41,* 103–146.

Polich, J., & Lawson, D. (1985). Event-related potentials paradigms using tin electrodes. *American Journal of EEG Technology, 25,* 187–192.

Popper, K. (1959). *The Logic of Scientific Discovery.* London: Hutchinson.

Potter, M. C. (1976). Short-term conceptual memory for pictures. *Journal of Experimental Psychology: Human Learning and Memory, 2,* 509–522.

Pritchard, W. S. (1981). Psychophysiology of P300. *Psychology Bulletin, 89,* 506–540.

Raymond, J. E., Shapiro, K. L., & Arnell, K. M. (1992). Temporary suppression of visual processing in an RSVP task: An attentional blink? *Journal of Experimental Psychology: Human Perception and Performance, 18,* 849–860.

Regan, D. (1989). *Human Brain Electrophysiology: Evoked Potentials and Evoked Magnetic Fields in Science and Medicine.* New York: Elsevier.

Ritter, W., Simson, R., Vaughan, H. G., & Friedman, D. (1979). A brain event related to the making of a sensory discrimination. *Science, 203,* 1358–1361.

Ritter, W., Vaughan, H. G. Jr., & Costa, L. D. (1968). Orienting and habituation to auditory stimuli: A study of short term changes in average evoked responses. *Electroencephalography and Clinical Neurophysiology, 25*, 550–556.

Rohrbaugh, J. W., Syndulko, K., & Lindsley, D. B. (1976). Brain wave components of the contingent negative variation in humans. *Science, 191*, 1055–1057.

Rosler, F., & Manzey, D. (1981). Principal components and varimax-rotated components in event-related potential research: Some remarks on their interpretation. *Biological Psychology, 13*, 3–26.

Rossion, B., Curran, T., & Gauthier, I. (2002). A defense of the subordinate-level expertise account for the N170 component. *Cognition, 85*, 189–196.

Rossion, B., Delvenne, J. F., Debatisse, D., Goffaux, V., Bruyer, R., Crommelinck, M., & Guerit, J. M. (1999). Spatio-temporal localization of the face inversion effect: An event-related potentials study. *Biological Psychology, 50*, 173–189.

Rossion, B., Gauthier, I., Goffaux, V., Tarr, M. J., & Crommelinck, M. (2002). Expertise training with novel objects leads to left-lateralized facelike electrophysiological responses. *Psychological Science, 13*, 250–257.

Ruchkin, D. S., & Wood, C. C. (1988). The measurement of event-related potentials. In T. W. Picton (Ed.), *Human Event Related Potentials* (pp. 121–137). Amsterdam: Elsevier.

Schendan, H. E., Ganis, G., & Kutas, M. (1998). Neurophysiological evidence for visual perceptual categorization of words and faces within 150 ms. *Psychophysiology, 35*, 240–251.

Scherg, M., Vajsar, J., & Picton, T. (1989). A source analysis of the human auditory evoked potentials. *Journal of Cognitive Neuroscience, 1*, 336–355.

Scherg, M., & von Cramon, D. (1985). A new interpretation of the generators of BAEP waves I–V: Results of a spatio-temporal dipole model. *Electroencephalography and Clinical Neurophysiology, 62*, 290–299.

Schmidt, D. M., George, J. S., & Wood, C. C. (1999). Bayesian inference applied to the electromagnetic inverse problem. *Human Brain Mapping, 7*, 195–212.

Schmolesky, M. T., Wang, Y., Hanes, D. P., Thompson, K. G., Leutgeb, S., Schall, J. D., & Leventhall, A. G. (1998). Signal timing across the macaque visual system. *Journal of Neurophysiology, 79*, 3272–3278.

Shapiro, K. L., Arnell, K. M., & Raymond, J. E. (1997). The attentional blink. *Trends in Cognitive Science, 1*, 291–296.

Shapiro, K. L., Raymond, J. E., & Arnell, K. M. (1994). Attention to visual pattern information produces the attentional blink in rapid serial visual presentation. *Journal of Experimental Psychology: Human Perception and Performance, 20*, 357–371.

Shibasaki, H. (1982). Movement-related cortical potentials. In *Evoked Potentials in Clinical Testing* (Vol. 3, pp. 471–482). Edinburgh: Churchill Livingstone.

Simson, R., Vaughan, H. G., & Ritter, W. (1977). The scalp topography of potentials in auditory and visual discrimination tasks. *Electroencephalography and Clinical Neurophysiology, 42*, 528–535.

Snyder, A. (1991). Dipole source localization in the study of EP generators: A critique. *Electroencephalography & Clinical Neurophysiology, 80*, 321–325.

Soltani, M., & Knight, R. T. (2000). Neural origins of the P300. *Critical Reviews in Neurobiology, 14*, 199–224.

Squires, K. C., & Donchin, E. (1976). Beyond averaging: The use of discriminant functions to recognize event related potentials elicited by single auditory stimuli. *Electroencephalography and Clinical Neurophysiology, 41*, 449–459.

Squires, N. K., Squires, K. C., & Hillyard, S. A. (1975). Two varieties of long-latency positive waves evoked by unpredictable auditory stimuli. *Electroencephalography and Clinical Neurophysiology, 38,* 387–401.

Sutton, S. (1969). The specification of psychological variables in average evoked potential experiments. In E. Donchin & D. B. Lindsley (Eds.), *Averaged Evoked Potentials: Methods, Results and Evaluations* (pp. 237–262). Washington, D.C.: U.S. Government Printing Office.

Sutton, S., Braren, M., Zubin, J., & John, E. R. (1965). Evoked potential correlates of stimulus uncertainty. *Science, 150,* 1187–1188.

Sutton, S., Tueting, P., Zubin, J., & John, E. R. (1967). Information delivery and the sensory evoked potential. *Science, 155,* 1436–1439.

Szücs, A. (1998). Applications of the spike density function in analysis of neuronal firing patterns. *Journal of Neuroscience Methods, 81,* 159–167.

Tallon-Baudry, C., Bertrand, O., Delpuech, C., & Pernier, J. (1996). Stimulus specificity of phase-locked and non-phase-locked 40 Hz visual responses in humans. *Journal of Neuroscience, 16,* 4240–4249.

Thorpe, S., Fize, D., & Marlot, C. (1996). Speed of processing in the human visual system. *Nature, 381,* 520–522.

Treisman, A. M. (1969). Strategies and models of selective attention. *Psychological Review, 76,* 282–299.

Tucker, D. M. (1993). Spatial sampling of head electrical fields: The geodesic sensor net. *Electroencephalography & Clinical Neurophysiology, 87,* 154–163.

Urbach, T. P., & Kutas, M. (2002). The intractability of scaling scalp distributions to infer neuroelectric sources. *Psychophysiology, 39,* 791–808.

Van Petten, C., & Kutas, M. (1987). Ambiguous words in context: An event-related potential analysis of the time course of meaning activation. *Journal of Memory & Language, 26,* 188–208.

van Schie, H. T., Mars, R. B., Coles, M. G., & Bekkering, H. (2004). Modulation of activity in medial frontal and motor cortices during error observation. *Nature Neuroscience, 7,* 549–554.

van Turennout, M., Hagoort, P., & Brown, C. M. (1998). Brain activity during speaking: From syntax to phonology in 40 milliseconds. *Science, 280,* 572–574.

Vasey, M. W., & Thayer, J. F. (1987). The continuing problem of false positives in repeated measures ANOVA in psychophysiology: A multivariate solution. *Psychophysiology, 24,* 479–486.

Vaughan, H. G., Jr. (1969). The relationship of brain activity to scalp recordings of event-related potentials. In E. Donchin & D. B. Lindsley (Eds.), *Average Evoked Potentials: Methods, Results and Evaluations* (pp. 45–75). Washington, D.C.: U.S. Government Printing Office.

Vaughan, H. G., Jr., Costa, L. D., & Ritter, W. (1968). Topography of the human motor potential. *Electroencephalography and Clinical Neurophysiology, 25,* 1–10.

Verleger, R. (1988). Event-related potentials and cognition: A critique of the context updating hypothesis and an alternative interpretation of P3. *Behavioral Brain Science, 11,* 343–427.

Verleger, R. (1997). On the utility of P3 latency as an index of mental chronometry. *Psychophysiology, 34,* 131–156.

Verleger, R., Gasser, T., & Moecks, J. (1982). Correction of EOG artifacts in event-related potentials of the EEG: Aspects of reliability and validity. *Psychophysiology, 19,* 472–480.

Verleger, R., Jaskowsi, P., & Wauschkuhn, B. (1994). Suspense and surprise: On the relationship between expectancies and P3. *Psychophysiology, 31*, 359–369.

Vitacco, D., Brandeis, D., Pascual-Marqui, R., & Martin, E. (2002). Correspondence of event-related potential tomography and functional magnetic resonance imaging during language processing. *Human Brain Mapping, 17*, 4–12.

Vogel, E. K., & Luck, S. J. (2000). The visual N1 component as an index of a discrimination process. *Psychophysiology, 37*, 190–123.

Vogel, E. K., & Luck, S. J. (2002). Delayed working memory consolidation during the attentional blink. *Psychonomic Bulletin & Review, 9*, 739–743.

Vogel, E. K., Luck, S. J., & Shapiro, K. L. (1998). Electrophysiological evidence for a post-perceptual locus of suppression during the attentional blink. *Journal of Experimental Psychology: Human Perception and Performance, 24*, 1656–1674.

Vogel, E. K., & Machizawa, M. G. (2004). Neural activity predicts individual differences in visual working memory capacity. *Nature, 428*, 748–751.

Wada, M. (1999). Measurement of olfactory threshold using an evoked potential technique. *Rhinology, 37*, 25–28.

Walter, W. G., Cooper, R., Aldridge, V. J., McCallum, W. C., & Winter, A. L. (1964). Contingent negative variation: An electric sign of sensorimotor association and expectancy in the human brain. *Nature, 203*, 380–384.

Wastell, D. G. (1977). Statistical detection of individual evoked responses: An evaluation of Woody's adaptive filter. *Electroencephalography & Clinical Neurophysiology, 42*, 835–839.

Winkler, I., Kishnerenko, E., Horvath, J., Ceponiene, R., Fellman, V., Huotilainen, M., Naatanen, R., & Sussman, E. (2003). Newborn infants can organize the auditory world. *Proceedings of the National Academy of Sciences, 100*, 11812–11815.

Woldorff, M. (1988). Adjacent response overlap during the ERP averaging process and a technique (Adjar) for its estimation and removal. *Psychophysiology, 25*, 490.

Woldorff, M. (1993). Distortion of ERP averages due to overlap from temporally adjacent ERPs: Analysis and correction. *Psychophysiology, 30*, 98–119.

Woldorff, M., & Hillyard, S. A. (1991). Modulation of early auditory processing during selective listening to rapidly presented tones. *Electroencephalography and Clinical Neurophysiology, 79*, 170–191.

Woldorff, M. G., Gallen, C. C., Hampson, S. A., Hillyard, S. A., Pantev, C., Sobel, D., & Bloom, F. E. (1993). Modulation of early sensory processing in human auditory cortex during auditory selective attention. *Proceedings of the National Academy of Science, 90*, 8722–8726.

Woldorff, M. G., Hackley, S. A., & Hillyard, S. A. (1991). The effects of channel-selective attention on the mismatch negativity wave elicited by deviant tones. *Psychophysiology, 28*, 30–42.

Wood, C. C., & McCarthy, G. (1984). Principal component analysis of event-related potentials: Simulation studies demonstrate misallocation of variance across components. *Electroencephalography and Clinical Neurophysiology, 59*, 249–260.

Woodman, G. F., & Luck, S. J. (2003). Serial deployment of attention during visual search. *Journal of Experimental Psychology: Human Perception and Performance, 29*, 121–138.

Woody, C. D. (1967). Characterization of an adaptive filter for the analysis of variable latency neuroelectric signals. *Medical and Biological Engineering, 5*, 539–553.

Index